Life of Fred®

Fractions

Life of Fred®
Fractions

Stanley F. Schmidt, Ph.D.

Polka Dot Publishing

ISBN: 978-0-9709995-9-7

Library of Congress Catalog Number: 2006936677
Printed and bound in the United States of America

Polka Dot Publishing Reno, Nevada

To order copies of books in the Life of Fred series,

visit our website PolkaDotPublishing.com

Questions or comments? Email the author at lifeoffred@yahoo.com

Twelfth printing

Life of Fred: Fractions was illustrated by the author with additional clip art furnished under license from Nova Development Corporation, which holds the copyright to that art.

for Goodness' sake

or as J.S. Bach—who was
never noted for his plain
English—often expressed it:

Ad Majorem Dei Gloriam

(to the greater glory of God)

If you happen to spot an error that the author, the publisher, and the printer missed, please let us know with an email to: lifeoffred@yahoo.com

As a reward, we'll email back to you a list of all the corrections that readers have reported.

A Note to Students

This is the story of one day in Fred's life. He's five years old, but he does some things that many fifty-five-year-olds have never done. Just turn to page 14 when you are ready to start reading about his adventures.

FOR NOW

When you read about what Fred is doing, go as fast as you like, but when you get to the math, please slow down. Math is more condensed than English. Most people have to read the math parts more than once in order to fully understand them. If you take your time, it will be enjoyable.

For now, put aside your calculators. Until you get to pre-algebra, one of the most important things you learn is your addition and multiplication facts. Adults who never learned what 7 × 8 equals are at a disadvantage.

Once you get to pre-algebra, you can take your calculator out of the drawer and use it all you like.

It is not necessary to get rid of your calculator. Just store it somewhere.

YOUR FUTURE

After this book, there are:

- ✯ Decimals and Percents
- ✯ Pre-Algebra
- ✯ Beginning Algebra
- ✯ Advanced Algebra
- ✯ Geometry
- ✯ Trigonometry and
- ✯ Calculus

after which, you are ready for university math courses as a junior* and may declare a major in mathematics as soon as your university allows you to.

---

* A junior at a university is a third-year college student.

A Note to Parents

You know what arithmetic books look like. They are all pretty much alike. Using very few words, they give a couple of examples and then have the students do a hundred identical problems. Then they give another couple of examples and another hundred problems. And for students, arithmetic becomes as much fun as cleaning up their rooms, eating yams, or going to the dentist.

Those authors often hope that they can fool their readers by throwing in a couple of irrelevant pictures of happy children at play.

Will these pictures make kids love math?

This book, *Life of Fred: Fractions*, takes a slightly* different approach. It tells a story—a story of one day in the life of a five-and-a-half-year-old boy. All of the math arises out of Fred's life. All of it is motivated—right down to when Fred (in Chapter 23) is working at the PieOne pizza place, and he's trying to decide whether to put the tomatoes on the pizza before or after it's cooked, and we get the commutative law.

FACTS ABOUT THE BOOK

Each chapter is a lesson. Thirty-two chapters = 32 lessons.

At the end of each chapter is a *Your Turn to Play*, which gives an opportunity for the student to work with the material just presented. The answers are all supplied. The questions are not all look-alike questions. Some of them require . . . thought!

✶ "Slightly" in the sense that fish swim slightly better than rocks.

In the Life of Fred series, your daily role in your children's math education is quite minimal. Just make sure that they are writing down the answers to the *Your Turn to Play* questions and encourage them not to cheat by just copying the answers out of the book. You do not need to check their *Your Turn to Play* answers.

As in all of the books in the *Life of Fred* series, the emphasis is on learning how to learn by reading. *Let the book do the math teaching.* Students of normal academic ability can learn mathematics from Fred without your tutoring the material. You can relax. As students progress through high school, college, and graduate school, they find that less and less is learned in the classroom lecture format. Increasingly, it's the written word that does the teaching. Things changed after Gutenberg.*

Learning how to learn by reading
is a very valuable skill.

translation:
Don't short circuit
learning to learn by reading
by your tutoring.

Your more active part in their math education occurs at the end of every four or five chapters. It is called **The Bridge**.

The Bridge consists of ten questions which review everything learned up to that point in the book. Under your supervision, they take out a piece of paper and write the answers to those ten questions. Then, together, you look at the answers that are given in the back of this book.

If they have answered 9 or 10 questions correctly, they have shown that they have *mastery* over the material and have earned the right to go on to the next chapter.

If they don't succeed on the first try, there is a second set of ten questions—a second try—for them to attempt. And a third try. And a fourth try. And a fifth try. Lots of chances to cross the bridge.

---

* Johannes Gutenberg figured out how to use movable type to print books. In 1455 he printed the Bible.

At the end of the book is **The Final Bridge**, fifteen questions. Again, there are five tries offered.

Life of Fred: Fractions covers a lot more than just how to add, subtract, multiply, and divide fractions. If you'll take a peek at the table of contents, which begins on the next page, you'll see how much is covered.

Have you ever wondered why, when you divide fractions,

$$\frac{2}{3} \div \frac{3}{4} \text{ becomes } \frac{2}{3} \times \frac{4}{3}\,?$$

Very few arithmetic books tell you *why*—they just say that it's a rule. Fred will give you the reasoning behind the rule.

I guess I should also mention: this book is very silly.

Contents

Chapter 1 Less Than 14
the symbol <
when to write in a book
the symbol >
Chapter 2 A Billion 18
a million
onomatopoetic words
minutes to hours to days to years
Chapter 3 Cardinal and Ordinal Numbers 21
inches to feet to yards
writing numbers out in words
the subjunctive mood
Chapter 4 Diameter and Radius 24
Fred's way of teaching
the meaning of $a < b < c$
Chapter 5 Fred's Budget 28
hyperbole
savings and expenses
The Bridge (five tries) following Chapter 5 31
Chapter 6 Doubling 37
doubling, and doubling, and doubling again
Chapter 7 Fractions 40
writing checks
the meaning of one-half and one-fourth
sectors
zero times any number
the symbol $\geq$
the symbol $\leq$
Chapter 8 Comparing Fractions 44
meridiem and ante meridiem
multiplying by ten
Chapter 9 Reducing Fractions 47
cutting an old comb into six equal pieces
Chapter 10 Add and Reduce 49
why you shouldn't run when you are carrying a thirteen-pound knife
earth to the sun in furlongs

The Bridge (five tries) following Chapter 10 52
Chapter 11 Subtracting Fractions with the Same Denominators 57
reading marks on a gauge
general rule: reduce fractions in your answer as much as possible
general rule: a fraction with zero on top is equal to zero
general rule: which fractions are equal to one
Chapter 12 Common Denominators 60
numerator
denominator
Chapter 13 Roman Numerals 62
numerals vs. numbers
when IIII is used
dividing CXLVI by XIV
Chapter 14 Adding Fractions 67
a 137-word sentence
bad eating habits
alliteration
Chapter 15 Fractions Mean Divide 71
for tapirs, $\frac{1}{2} + \frac{1}{2}$ does not equal one
whole numbers
the integers
the imaginary numbers
The Bridge (five tries) following Chapter 15 74
Chapter 16 Least Common Multiple 79
selecting a common denominator
Chapter 17 Improper Fractions 81
least common denominator
improper fractions
mixed numbers
Chapter 18 Lines of Symmetry 84
equilateral triangles
Chapter 19 Division by Zero 87
The Bridge (five tries) following Chapter 19 90
Chapter 20 Subtracting Fractions 95
lb. = pound
Chapter 21 Circumference 98
multiplying fractions
Chapter 22 Multiplying Mixed Numbers 100
changing mixed numbers into improper fractions
Chapter 23 Commutative Law 103

Chapter 24 Adding Mixed Numbers. 105
The Bridge (five tries) following Chapter 24. 108
Chapter 25 Canceling. 113
Florence Nightingale
of often means multiply
Chapter 26 Opposites. 115
definition of a function
inverse functions
Chapter 27 Area of a Rectangle.. 117
goal posts: high school vs. college
Chapter 28 Unit Analysis. 120
rules for kitties
dimensional analysis
square of a number
square roots
Chapter 29 Subtracting Mixed Numbers. 123
borrowing
The Bridge (five tries) following Chapter 29. 126
Chapter 30 Division by a Fraction.. 131
credit card applications
how to divide by a fraction
why you invert and multiply
Chapter 31 Geometry. 137
Boyle's law of gases
diagonals of a plane figure
vertex of a plane figure
right angles
parallel lines
Chapter 32 Estimating Answers. 140
The Final Bridge (five tries) concluding the book. 142
Answers to all the Bridge Problems in the book. 147
Complete Solutions to all the *Your Turn to Play*. 166
Index.. 191

Chapter One
Less Than

When Fred first arrived at KITTENS University he could barely walk. That was because he was only nine months old. A student named Betty became his friend, and she would often carry little Fred so that he could get to class on time.

But that was many years ago. Fred is now 5½ years old. He's no longer a baby who needs to be carried. He walks to class.

Fred walking

Now that he's 5½ years old, he sees things that he never noticed before. He notices that a lot of students ride bikes. They can go fast. They get to wear helmets, and that looks cool.

Fred thinks to himself *I want a bike!*

When he was only five years old, he was very happy just walking. But now that he is 5½, he is older.

Fred stopped and took a piece of paper out of his pocket. He started making a list.

Why I Want a Bike

1. I can get to class faster.
2. When I'm on a bike, I am taller.
3. I get to wear a helmet. It would look silly to wear a helmet if I'm just walking.
4. I will need a lock. Locks are fun.

Fred took out a second piece of paper and continued his list.

Why I Want a Bike

page 2

5. I am no longer a baby. I used to be 5, but now I'm 5½.

5 < 5½

Wait! Stop! You, my reader, shout. **I don't get it. What does 5 < 5½ mean?**

This is a very silly book. This is a book in which you, the reader, can talk back to me, the author. When you are talking to me you use ***this type font.*** When I am telling the story I use Times New Roman type font. When Fred is thinking, he uses this type font.

Fred knows a lot of math. Sometimes he uses math symbols instead of English words. Instead of writing "Five is less than five and a half," he wrote "5 < 5½."

The symbol "<" means "is less than."

For example, 3 < 8
20 < 40
0 < 12 and
4 < 390724902304237230049.

It's NOT true that 9 < 7.

Now let's continue watching Fred make his list.

Why I Want a Bike

page 2

5. I am no longer a baby. I used to be 5, but now I'm $5\frac{1}{2}$.

   $5 < 5\frac{1}{2}$

6. I can walk at 3 mph. On a bike I can go 10 mph. Everyone knows that $3 < 10$. Riding is faster.

Before you yell ***stop!*** again . . .

mph means miles per hour

Now it is your turn, my reader, to do some writing. Please get out a piece of paper. When you were a baby, you may have had books that you wrote in. Those workbooks gave you a problem like 2 + 3 = ____ and you would write in the book: 2 + 3 = __5__.

You are no longer a baby. If you write in this book, you will mess it up for any younger brothers or sisters who want to read it.

The rule for writing in books—and elsewhere—could be very complicated:

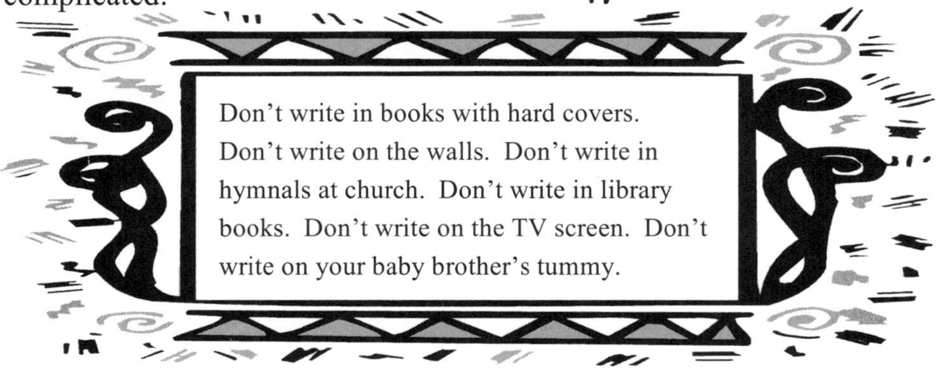

Instead, here is an easy rule:

Write only in books you bought with your own money.

Do you have your piece of paper yet?

At the end of every chapter in this book is *Your Turn to Play*. It is a chance for you to write.

There are three important ways that people learn: reading, hearing, and writing.

Just silently reading the math or just hearing someone read it aloud is not enough.

Your Turn to Play gives you a chance to learn by writing.

Your Turn to Play

1. On your paper, write the words that finish this sentence:
The symbol < stands for. . .

2. Is $88 < 92$ true or false?

3. Is $100 < 12$ true or false?

4. Is $5 < 5\frac{1}{2}$ true or false? ☺☺☺☺☺ $\overset{?}{<}$ ☺☺☺☺☺☺

5. Fill in any number that makes this true: $14 <$ __?__.

6. Fill in any number that makes this true: __?__ < 3.

7. Add $389 + 772$.

8. Make a guess. If < means "is less than," what does > mean?

Complete solutions are on page 166

Have you ever heard the saying: *The pen is mightier than the sword*? This means that the written word is more powerful than a physical weapon. The saying first appeared in its present form in the play *Richelieu* in 1839:

> *Beneath the rule of men entirely great,*
> *The pen is mightier than the sword.* (Act II)

But many people expressed that idea before 1839:

"The tongue is mightier than the blade." —Greek poet Euripides in the 400s B.C.
". . . many wearing rapiers are afraid of goose quills." —Shakespeare in *Hamlet* in 1600.

In mathematics, we might express it: **Pen > Sword**

Fred remembered Toad in *The Wind in the Willows.* In that book, Toad went nuts about cars. Toad thought about nothing else. Toad *needed* a car.

Fred was more grown up than any old frog. He didn't *need* a bike, he just *wanted* one. So to prove that he wasn't a baby (or a toad), Fred took out another piece of paper and made a list of the drawbacks of owning a bike.

Why I Shouldn't Get
a Bike

1.

I can't think of any reasons Fred thought to himself. Bikes are so cool. I could get to class faster. I could feel the wind in my hair as I rode.

Clearly, Fred was in love with the idea of owning a bike. And when you are in love, you don't think very well. Fred was telling himself "I could feel the wind in my hair as I rode." How silly! Look at Fred. Do you see any hair?

And even if he did have a lot of hair, he would be wearing a helmet.

So Fred had six reasons why he should have a bike and zero reasons why he shouldn't. 6 > 0.

As Fred continued to walk to class, three students on bikes passed him. Zoom! Swish! Whoosh!* And each time, Fred thought to himself I want a bike!

"Hi! Fred," Betty called out. "Where are you going? You just passed the classroom."

"Oh! Did I?" Fred answered. He blushed a little bit. "I guess I was lost in thought."

Betty said, "I bet you were thinking about math."

"No," answered Fred. "This time I wasn't. I was thinking about getting a bike. I need—I mean, I want a bike." He didn't want to sound like that Toad in *The Wind in the Willows*. "I can think of six reasons why I should get a bike, and I can't think of any reasons why I shouldn't."

If this book were printed in color, we would add a little pink to Fred's face.

"I bet," Betty said, "that if you thought hard enough, you could think of a billion reasons to own a bike."

"A billion is 1,000,000,000—one followed by nine zeros," Fred said. "That's a lot of reasons. A billion is a thousand million: 1,000 × 1,000,000."

You could sense that he was warming up the math part of his brain.

---

* This is called a footnote. It explains something that needs explaining, but without having to stop the story. That's why we hide it down here at the bottom of the page.

What needs explaining are words like *zoom*, *swish*, and *whoosh*. These are onomatopoetic words [ON-oh-mat-eh-poe-ET-ik]. Seven syllables! Onomatopoetic words are words that sound like the thing that they are describing. When a bike passes close to Fred, it sounds like *swish*.

When a bee buzzes, it sounds like *buzzzzzzz*. *Buzz* is an onomatopoetic word.

A snake hisses. A prison door swings shut. Clank!

Comic books are loaded with examples of onomatopoeia. That word has only six syllables [ON-oh-mat-eh-PEE-a]. I much prefer the word *onomatopoetic*. It's longer.

Of course, if you really want to get silly, you could say that a person who uses lots of words like *boom, fizz,* or *crackle* enjoys speaking onomatopoetically. That's nine syllables.

"I wonder how long it would take to list a billion reasons for me to own a bike?" Fred continued. "Let's say I list one reason every second. Then I could list 60 reasons every minute, since there are 60 seconds in a minute."

Your Turn to Play

1. If Fred could list 60 reasons to own a bike every minute, how many could he list in an hour?

2. How many could he list in a 24-hour day?

3. Since 86,400 is nowhere near a billion, let's see how many reasons Fred could list in a year. We'll assume that there are 365 days in a year. (Actually, there are about 365¼ days in a year. Because a year isn't exactly 365 days, we have leap years in which we add an extra day to February. And because a year isn't exactly 365¼ days—it's closer to 365.242199 days—we have leap seconds!)

4. So, working day and night for a whole year and listing one reason each second why he should own a bike, Fred would have listed only 31,536,000 reasons. 31,536,000 < 1,000,000,000. Would he be able to list a billion reasons if he spent thirty-one years working?

Complete solutions are on page 166

If you look at Fred's check on page 167, you might notice several things. First of all, his last name is Gauss. That rhymes with *house.* It also rhymes with *mouse* and with *louse.* A louse is a little flat insect that likes to bite people.

When Fred was in kindergarten, some of the other kids would tease him and call him Fred Mouse or Fred Louse. That wasn't very nice.*

The other thing to notice on the check is Fred's address. It is room 314 in the Math Building at KITTENS University. In many buildings, rooms in the one hundreds (such as room 125 or room 193) are on the first floor. Rooms in the two hundreds (such as room 280) are on the second floor. Since Fred's room is room 314, it would be a good guess that he is on the third floor. If the Math Building were very tall, room 1425 would be on the fourteenth (14th) floor.

You know two different ways to count. If you take a bunch of pennies out of your pocket and count them, you would say, "One, two, three, four." These are **cardinal numbers**—counting numbers. Six forks. Zero mice. But if you saw this line of people all waiting to get an ice cream cone, you would say that Fred is fourth.

First, second, third, fourth, fifth are **ordinal numbers**. With ordinal numbers the order is important.

ordinal ↔ order

✶ The word *nice* reminds me of the plural forms of *mouse* and *louse*. One mouse, two mice. One louse, two lice. But the plural of *house* isn't "*hice*." One house, two houses. English can be very strange.

But Fred wasn't waiting in line for an ice cream cone. He was heading to class at KITTENS University. And now, with his head full of thoughts of owning a bike, he entered the classroom.

There were hundreds of students in the classroom. When Fred walked up to the front of the room, everyone looked at him. He was about to start teaching.

Wait! Stop! I, your reader, need to ask something. You say, "He was about to start teaching." This boy is only five and a half years old. Where is the teacher of this class?

Good question. I think I forgot to mention that Fred *is* the teacher.

That's not possible. Fred should be in kindergarten at that age.

Well, Fred is not your average five-and-a-half-year old. For one thing, he is pretty short. Did you see the picture of him and other kids on the previous page? He is only 36 inches tall. (That can be written as 36".) Most kids who are five and a half years old are a lot taller than 36".*

I, your reader, was asking about why he was a teacher at the university. I wasn't asking about his height. Tell me! Why isn't this kid in kindergarten?

As I was saying, Fred is not your average five-and-a-half-year old. He did a lot of reading at home in his early years and only spent several weeks in kindergarten. The complete story of the first five years of his life is told in *Life of Fred: Calculus.* You'll have to wait until you get to calculus to read about that part of his life.

Wait! I don't like waiting. How long do I have to wait? And what is calculus?

So many questions! Getting to calculus won't take that long. First you study fractions—which you are doing now. Then decimals and percents. Then pre-algebra. Then beginning algebra,

advanced algebra,

geometry,

trig, and then

calculus.

✶ Since there are 12 inches in a foot, 36" = 3 × 12". So Fred is three feet tall. Three feet can be written as 3'.

Since there are three feet in a yard, Fred is one yard tall.

Fred read about those things—fractions, decimals, percents, pre-algebra, beginning algebra, advanced algebra, geometry, trig, and calculus—at home. Now he teaches them.

Okay. Please get on with the story. You say he walked up to the front of the classroom and is ready to teach.

I can't go on. We're at the end of the chapter.

Your Turn to Play

1. Is one billion a cardinal number?
2. Name a very small cardinal number.
3. 48 inches is how many feet?
4. 100" is how many feet? (100" means 100 inches.)
5. If you ask, "How many floors does the Math Building have?" will the answer be a cardinal number or an ordinal number?
6. If you ask, "Which floor is Fred's office on?" will the answer be a cardinal number or an ordinal number?
7. Fred is three feet tall. If he were 17' tall, how much taller would he be than his current height? (17' means 17 feet.)
8. If you have $7, how much more money would you need in order to have a billion dollars?
9. Write out your answer to the previous question in words.

Complete solutions are on page 167

Chapter Four

Diameter and Radius

Fred stood at the front of the classroom. In addition to the hundreds of students, there were two TV cameramen, five reporters, several photographers, an artist, and three people with laptops who were writing blogs about Fred as he taught.

Someone from a radio station put a microphone in front of Fred so the world could hear his words. A sculptor had set up a little studio in the back of the room and was busy making a statue.

In short, Fred was a very popular teacher.

Everyone applauded. "Clap! Clap! Clap!"* A dog barked. The dog was taken outside, and everyone became quiet. The sculptor stopped hammering.

Everybody liked the way that Fred taught math. One of the magazines had called it "The Fred Way." The old way of teaching math was dull and dry. Using the old way, a teacher would write on the board $\frac{2}{9} + \frac{3}{9} = \frac{5}{9}$ and then would give the students 500 problems to do for homework.

The Fred Way was new, but it was also as old as the Greeks and the Bible. Fred would tell stories. Everyone likes to hear a good story, and the math would just be a natural part of the story.

Using the old way, the math would be pounded into the student's head—like the homework pictured on the right side of this page.

Using the Fred Way, the math was giggled into place and was too funny to forget.

the old boring way

Your Homework

1. 4/7 + 2/7 = ?
2. 8/11 + 1/11 = ?
3. 40/51 + 2/51 = ?
4. 3/10 + 4/10 = ?
5. 5/66 + 2/66 = ?
6. 19/41 + 2/41 = ?
7. 33/67 + 33/67 = ?
8. 9/28 + 18/28 = ?
9. 34/667 + 50/667 = ?
10. 1/3 + 1/3 = ?
11. 13/27 + 13/27 = ?
12. 5/81 + 55/81 = ?
13. 17/44 + 10/44 = ?
14. 1/7 + 1/7 = ?
15. 2/7 + 1/7 = ?
16. 4/889 + 98/889 = ?
17. 98/777 + 4/777 = ?
18. 38/91 + 6/91 = ?

honors problems:

19. 3/19 + 2/19 + 5/19 = ?
20. etc., etc., etc.

✶ A little onomatopoeia.

"Once upon a time," Fred began, "a four-year-old girl named Meddie Mittens decided that she wanted a bike. She had seen all her friends with bikes, and she wanted one too."

The photographers snapped pictures. Some of the students started taking notes. The sculptor held his hammer, waiting for Fred to take a break in his lecture so that he could start hammering again.

What a blogger wrote

"So Meddie went to her mom and announced, 'Mom, I need a bike.' Her mother, Cheryl, smiled and explained to Meddie about a toad in *The Wind in the Willows.* Meddie didn't understand what Cheryl was talking about. Meddie reiterated (repeated) her request, 'Mom, I must have a bike! I gotta have one or I'll die.' Meddie made a very sad face."

The artist squeezed out some oil paint and began her painting.

Fred continued, "So Cheryl gave her daughter a big hug and said, 'I think you will have to talk to your daddy. We don't have a lot of money right now. Let's see what he says when he gets home.'

"Meddie Mittens exclaimed, 'I can think of six reasons why I need a bike.' "

One of the students in the class shouted out, "Cardinal number!"

"Cheryl asked, 'And what is your first reason?' "

Another student in the class shouted, "Ordinal number!"

"Meddie thought for a moment and said, 'I want to get a helmet with a flower on it, and I need a bike first so that you will have to buy me a helmet.' "

Fred looked at the clock on the wall and announced, "Our time is up for today. I'll continue the story tomorrow."

Everyone applauded again. The TV cameramen turned off their cameras. The artist put away her oil paints. The people left the classroom, all except for Joe and Darlene. Joe was still working on his class notes.

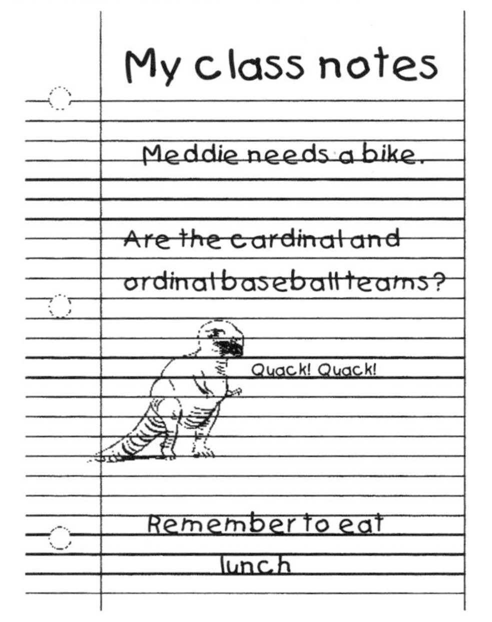

"Are you ready yet?" Darlene asked Joe. Joe always took a long time to write his class notes.

"Did I ever tell you about the bike I owned?" Joe asked. "It was red and blue and had 18" wheels."

"Eighteen-inch wheels," Darlene said. "Was the diameter equal to 18" or was the radius equal to 18"?"

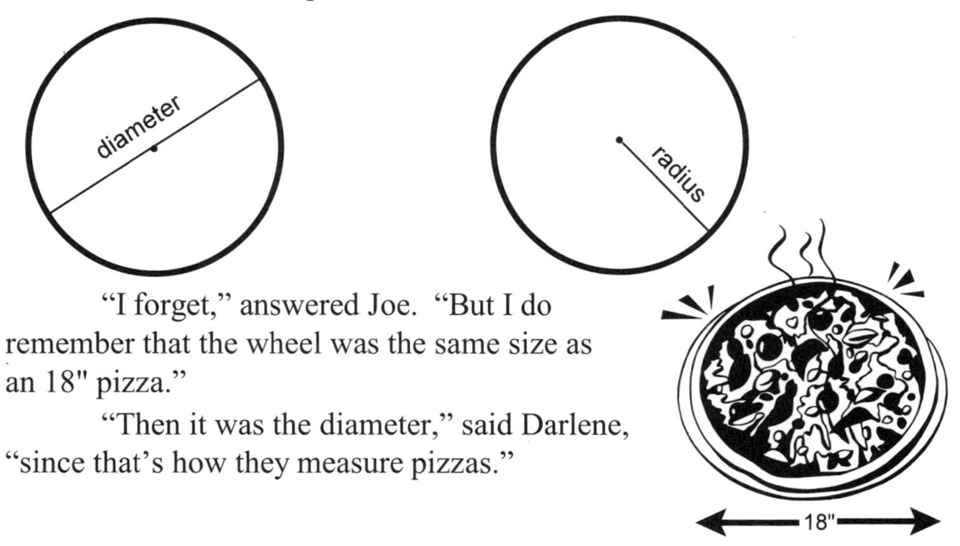

"I forget," answered Joe. "But I do remember that the wheel was the same size as an 18" pizza."

"Then it was the diameter," said Darlene, "since that's how they measure pizzas."

Your Turn to Play

PLEASE REMEMBER TO TAKE OUT A PIECE OF PAPER AND WRITE YOUR ANSWERS. PLEASE DON'T JUST READ THE QUESTIONS AND READ THE ANSWERS. YOU'LL LEARN MORE BY WRITING.

1. If the radius of a pizza is 17", what is its diameter?

2. Convert 34" into feet and inches.

3. During Fred's lecture, Joe was eating jelly beans. He ate sixty-two thousand, eight hundred four of them. (They were small.) Write that in numbers. Is that a cardinal or an ordinal number?

4. Joe had one jelly bean that he hadn't eaten. He put it on the floor and smashed it with his foot. "Look!" he said to Darlene. "It looks like a little green pizza." (Joe had gotten full and was playing with his food.) He picked it up and said, "If I were hungry [subjunctive mood], I would eat it."

"If you were three times brighter," said Darlene, "you would have an IQ of 150." What was she saying Joe's actual IQ is?

5. $8 < 11 < 15$ means that 8 is less than 11 *and* that 11 is less than 15. Name a number—let's call it x—so that $4 < x < 6$.

Complete solutions are on page 168

Chapter Five
Fred's Budget

After the class was over, a reporter from the student newspaper stopped Fred. She put a microphone in front of him and asked him if he would answer some questions.

"I would be happy to," he replied.

"In your lecture you said that Meddie really wanted a bicycle or she would die. Did she really want a bike that much?"

Fred thought for a moment and then answered, "Of course, Meddie is only four years old. Just a child. Kids that age tend to use a lot of hyperbole."*

"You also said that Meddie could think of six reasons to own a bike. Was that also hyperbole? How could someone have that many reasons?"

Fred blushed a little. He didn't know how to answer that question. He thought It's easy to think of six reasons to own a bike. Before class Betty kidded me and said that I could think of a billion reasons. If you really want something, you can always think of six reasons.

Before Fred could answer the reporter's question, she asked another. "And what about Mr. and Mrs. Mittens not having enough money to buy a bike for their kid?"

Fred couldn't answer that question either. He wasn't sure what she was really asking. Was she asking me:

1. Why didn't they have enough money? or was she asking
2. Was Cheryl lying to her daughter when she said they didn't have enough money? or was she asking
3. Did they really not love their kid?

"If you remember my lecture," Fred answered, "I didn't say that they didn't have enough money. All I said was that Cheryl told her daughter that 'We don't have a lot of money right now.' "

---

* Hyperbole [high-PURR-beh-lee]. Exaggeration. Mothers use it when they say, "I've told you a million times to clean up your room!" Darlene would be using hyperbole if she tells Joe, "It takes you forever to write your class notes."

The reporter thanked Fred for his time and hurried off to write up the story for the school newspaper.

Fred headed back to the Math Building. He climbed the stairs to the third floor (ordinal number) and went into room 314. This was his office at the university. It was also where he lived. Every night, he slept under his desk in his office. He was paid $500 per month for teaching at the university. Like Mr. and Mrs. Mittens, he didn't have a lot of money, so he saved money by not renting an apartment or buying a house.

There was a knock on the door.

"Come in."

It was Betty. She said, "Have you seen the school newspaper? It has just come out." She showed him a copy.

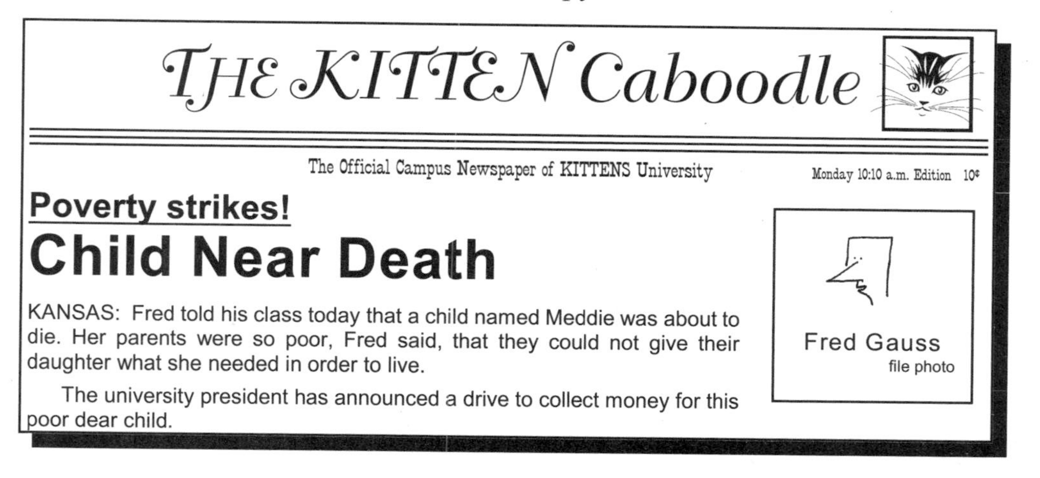

THE KITTEN Caboodle

The Official Campus Newspaper of KITTENS University

Monday 10:10 a.m. Edition 10¢

Poverty strikes!

Child Near Death

KANSAS: Fred told his class today that a child named Meddie was about to die. Her parents were so poor, Fred said, that they could not give their daughter what she needed in order to live.

The university president has announced a drive to collect money for this poor dear child.

Fred Gauss
file photo

Wait! Stop! I, your reader, don't understand. Fred didn't tell his class that Meddie was dying. I don't get it.

Sometimes, newspapers make mistakes. Sometimes they make lots of mistakes. It was true that the university president was starting a drive to collect money for Meddie. But that was a mistake that the president was making. He thought that Meddie was a real person, rather than just being part of the story that Fred was telling to his class.

At least the newspaper got the picture of Fred right. It would have really been a mistake if they printed this →

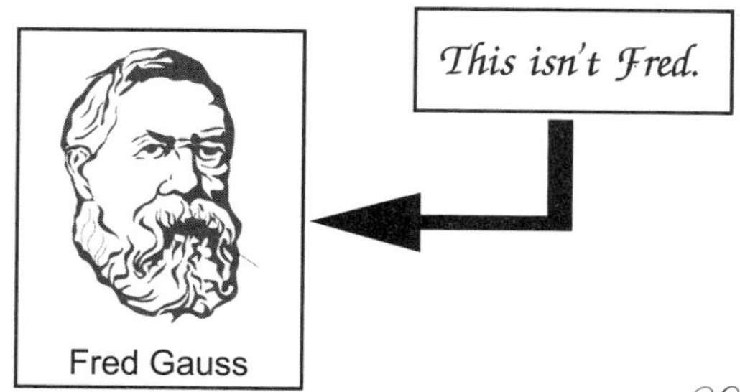

"How much does a bike cost?" Fred asked Betty. "I make $500 each month teaching here at KITTENS University. In a year that would be twelve times as much."

On the blackboard in his office he wrote:

$$\begin{array}{r} 500 \\ \times\ \underline{\ 12} \\ 1000 \\ \underline{500\ \ } \\ 6000 \end{array}$$

"So I make $6,000 per year. Is that enough money to buy a bike? I don't know. I've never bought a bike before."

Betty smiled. "It depends on what kind of bike you want to buy. Some cost a lot, and some are pretty cheap. But what you *earn* is not the same as what you *save*. At the end of a year, you won't have $6,000. What are your expenses each month?"

Your Turn to Play

1. Fred told Betty that each month he spent $6 for clothes, $0 for housing, $26 for food, $50 for Sunday school offering, $30 for book purchases, and $2 for miscellaneous things. What was the total amount Fred was spending each month?

2. If Fred makes $500 per month, how much does he save each month? (You'll need the answer to problem 1, above, to figure out how much he saves.)

3. What is half of $500?

4. Does Fred save more or less than half his income?

5. From the previous problems, you know how much Fred saves each month. How much does he save each year?

6. If Fred's income and expenses were to stay the same each year, how much would Fred save in 10 years?

7. How much would he save in 100 years?

Complete solutions are on page 169

Wait! Stop! I, your reader, think you made a mistake. What's this?

I was trying to draw a bridge.

I know it's a bridge. But what's a bridge doing in the middle of Fred's office? Fred was reading a newspaper that Betty brought him. Then they started to talk about how much money Fred could save. Mr. Author, you can't do that!

Calm down. Please let me explain.

Okay. Start explaining.

We were in the middle of Fred's office.

I know that! I just told you that!

Please. I'm explaining as fast as I can.

But I was just telling you that you can't draw bridges in the middle of Chapter 5 in Fred's office.

This book is going to be a million pages long* if you keep interrupting me. Here are the facts: We aren't in the middle of Chapter 5. You just finished the *Your Turn to Play* for Chapter 5.

So we're in Chapter 6.

No we're not. We are at **The Bridge**.

After every four or five chapters, we give you the chance to show that you haven't forgotten what you've learned. After you prove that you remember, then you head on to the next chapter.

Okay. I'm ready. Bring it on. Where's this bridge thing?

Please take out a piece of paper and a pencil (or pen), and turn the page.

---

* hyperbole

The Bridge

from Chapters 1–5 to Chapter 6

first try

Goal: Get 9 or more right and you cross the bridge.

After you have written the answers to all ten problems, and after you have checked your work as much as you want to, it will be time to find out whether you have crossed the bridge into Chapter 6. The answers are in the back of this book starting on page 147. Good luck!

1. When Fred turns 6 years old, KITTENS University will pay him $600 each month. How much would he make in a year?
2. If the radius of a circle is 53", what is the diameter?
3. A tree was 17 feet tall. Years later it was 233 feet tall. How many feet did it grow?
4. Mark each of these as True or False:
   $7 < 98$
   $987 > 900$
   $55 > 555$
5. Write out in words: 4,345,211.
6. Add 893 + 237.
7. How many minutes are there in 24 hours?
8. Fred gave a test to his math class. Joe came in 93rd. Is this a cardinal or an ordinal number?
9. 200" is how many feet? Express your answer in feet and inches.
10. If you had a million dollars, how much more money would you need to have a billion dollars?

If you got nine or ten correct, you are ready to go on to Chapter 6, which begins on page 37.

If not, then you probably have been reading too fast. (Remember the story of the tortoise and the hare. The speedy one is not always the one who finishes first.) You may earn the right to another try by first correcting all the errors you made. Then—and only then—may you attempt the bridge again. Please see the next page for a second try.

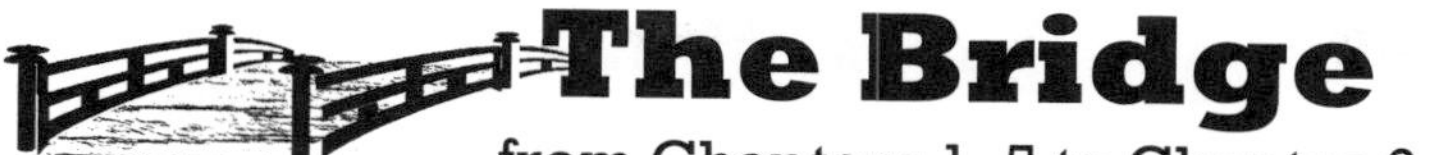

The Bridge

from Chapters 1–5 to Chapter 6

second try

1. Fill in any number that makes this true: 55 > __?__
2. Write as a number: three million, eight hundred sixty-four thousand, nine hundred ninety-nine.
3. If the radius of a circle is 39", what is its diameter?
4. Convert 38" into feet. Express your answer in feet and inches.
5. Add 593 + 498.
6. Divide 449 by 13.
7. Name a number—let's call it y—so that 7 < y < 10.

8. This week Darlene has $18. Next week she will have $302. How much will she have gained in a week?

9. What is one-third of 180?
10. Write out in words: 4,000,000,017.

After you have written the answers to all ten problems, and after you have checked your work as much as you want to, it will be time to find out whether you have crossed the bridge into Chapter 6 this time. Good luck!

If you got nine or ten correct, you are now ready to go on to Chapter 6.

If not, then correct the errors you've made on this second try, and you will have earned the right to a third try. Please see the next page.

The Bridge

from Chapters 1–5 to Chapter 6

third try

---

1. What is the ordinal number that comes right after seventeenth?
2. This is the third try for this bridge. Is *third* a cardinal number?
3. Here is what Joe spends each month: $400 for rent, $3 for presents for Darlene, $38 for jelly beans, and $5 for miscellaneous things. What is the total amount Joe spends each month?

Joe's Jelly Beans

4. 300" is how many feet?
5. Mark each of these as True or False:
   - 7 > 99
   - a billion > a thousand
   - 8 < 2
6. If you had three million dollars, how much more money would you need to have a billion dollars?
7. If a circle has a diameter equal to 88 feet, what would its radius be? (Remember to write "feet" as part of your answer.)
8. If Fred made $800 each month, how much would he make in a year?
9. Write as a number: seven billion, three hundred sixty-six.
10. Add 963 + 877.

You know how it works now. If you have nine or more correct, you may go on to Chapter 6. If not, then correct your errors, and you will have earned the right to try again. Your fourth try (ordinal number) is on the next page.

The Bridge

from Chapters 1–5 to Chapter 6

fourth try

---

1. Write out in words: 2,000,600,000.

2. What is one-third of 210?

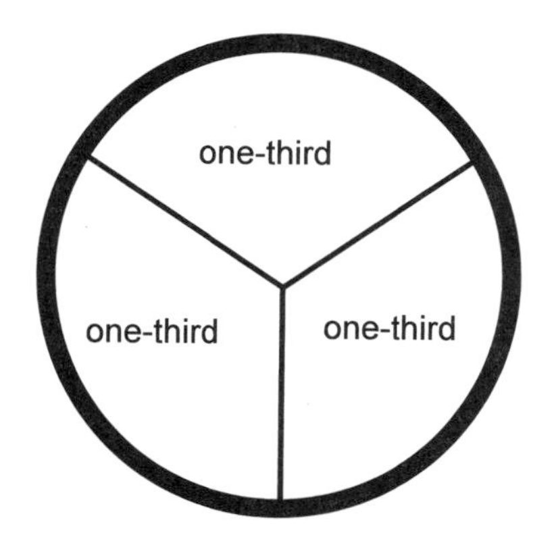

3. Add 792 + 798.
4. What is the cardinal whole number that comes right after one million? (The whole numbers are 0, 1, 2, 3, 4, 5, 6, 7, 8, 9, 10, 11, 12, 13, 14. . . .)
5. If the radius of a pizza is 8", what is its diameter?
6. Name a number—let's call it w—so that $50 < w < 100$.
7. 88" is how many feet? Express your answer in feet and inches.
8. Mark each of these as True or False:
   $0 < 10$
   $23 > 14$
   a billion > 0
9. Here is what Darlene spends each month: $500 for rent, $77 for presents for Joe, $49 for clothes, and $9 for miscellaneous things. What is the total amount she spends each month?
10. How many minutes are in 39 hours?

Did you get nine or more correct? If so, you can head on to Chapter 6. You've earned it. Otherwise, please correct your errors and try again on the next page. It doesn't matter whether you learn the math quickly or slowly. It really doesn't. The only thing that counts is whether you learn it. It is not a race.

The Bridge

from Chapters 1–5 to Chapter 6

fifth try

---

1. If you had a million dollars, how much more money would you need to have $3,000,007?
2. Fill in any number that makes this true: 10 < __?__.
3. Write as a number: eight million, two hundred eighty-eight thousand, one hundred six.
4. If Fred made $900 each month, how much would he make in a year?
5. What is one-third of 240?
6. This is the fifth try for this bridge. Is *fifth* an ordinal number?
7. If a circle has a diameter equal to 50 miles, what would its radius be? (Remember to write "miles" in your answer.)
8. Write out in words: 2,000,827.
9. My horse is 59 inches tall. Your horse is 82 inches tall. How much taller is your horse than mine?

My horse

Your horse

10. Divide 388 by 7.

If you got nine or ten right, turn to the next page to start Chapter 6. If not, then please correct your errors and turn back four pages to "first try."

It was time to buy a bike. Fred still didn't know how much they cost. He hoped that they wouldn't cost too much. Betty suggested that they look in the school newspaper to see if there were any ads for bikes. Fred turned to page two.

THE KITTEN Caboodle

The Official Campus Newspaper of KITTENS University — Thursday 10:10 a.m. Edition 10¢

Page Two

Help Wanted
PieOne Pizza

We need someone
to make the pizzas ⇒ pizza cook
to serve the pizzas ⇒ server
to take their money ⇒ cashier

Take books out for free at the KITTENS University library.

Cheap Bikes!
Just fell off the back of a truck. Can sell them really cheaply.

C. C. Coalback
bicycles
and other "found" stuff

There it was! A bicycle store. And they have cheap bikes. Fred got very excited. He was almost as excited as the time he found out that you can take books out of the library for free.

"Betty, I've got to go to the bike store," Fred said. Before Betty could say bye, he ran

and six blocks east and

turned and there it was.

Fred walked into the store and said, “I need a bike.” He didn’t say that he wanted a bike or that he was looking for a bike. He said that he *needed* a bike.

Mr. Coalback smiled and walked over to a bicycle and changed the price.

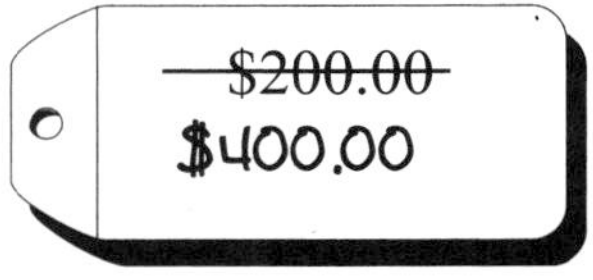

“You say you need a bike,” Coalback responded. “I have a very nice bicycle here. It has a very special price. Do you want to pay cash, write a check, or put it on your credit card?”

“Could I look at it first?” Fred asked.

“You’ll have years to look at it after you’ve bought it,” Coalback told Fred. “Why waste your time and waste my time?”

“Well. I have never shopped for a bike before. I don’t know anything about them or what they cost. Could I please see it before I buy it?”

Coalback went over and changed the price tag again. Coalback figured that if Fred *needed* a a bike and if Fred really didn’t know how much bikes cost, he could charge Fred a lot of money. That’s why he kept doubling the price.

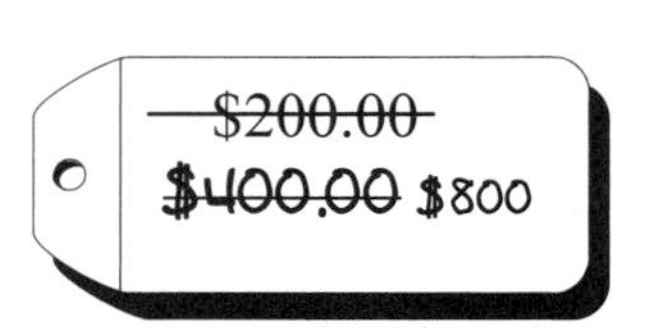

“Okay. If you really want to see it before you buy it, come with me,” Coalback said.

“I’m so excited. I can hardly wait to take it home. I can show it to my friends. I will ride it everywhere I go.” Although Fred didn’t know it, he was telling Coalback that he was going to buy a bicycle regardless of how much it cost. And Coalback knew exactly what to do. He doubled the price again.

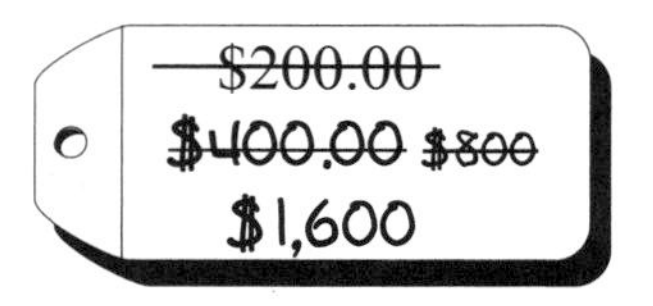

Your Turn to Play

1. The price of the bike is now $1,600. If Coalback doubles it again, it will be $3,200. How many times will Coalback have to double the price before it is over a million dollars?

2. Back in Chapter 5, we found that Fred saves $386 each month. Could he save enough in four months to buy a $1,600 bike?

3. Could he save enough in five months?

4. If a bicycle wheel has a radius of 21", what would its diameter be?

Complete solutions are on page 170

Chapter Seven
Fractions

Coalback led Fred to the back of the store. "Here it is," Coalback told Fred. "It's the bicycle of your dreams. Ain't she pretty? And it's all wrapped up ready to go. And the best part is the price. It's only $1,600."

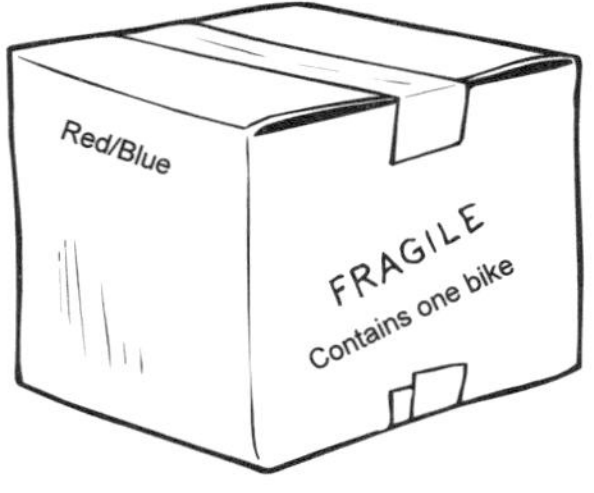

Fred gulped. He thought to himself *That's almost five months of savings for me. But I really need a bike.* Fred nodded. He was going to buy it. He took out his Kittens Bank™ checkbook.

Coalback said, "If you'll excuse me for a moment, I need to check on something in the back room." He dashed into the back room and phoned a friend who worked at the bank. Here is a tape recording of their conversation:

> Coalback: "Let me speak to Lester Snow."*
> Snow: "Yeah. This is him."**
> Coalback: "I've got another sucker here. How much does he have in his checking account?"
> Snow: "What's his name?"
> Coalback: "I don't know. I'll be back in a minute."

Coalback rushed back out and said, "Hey kid. What's your name?"

Fred smiled. He was happy that Mr. Coalback was taking such a personal interest in him. "My name is Fred Gauss. It rhymes with house."

Coalback dashed into the back room again.

> Coalback: "The kid says his name is Fred Gauss. Ever heard of him?"

✶ It's more polite to say, "May I please speak to Lester Snow."

✶✶ I know you wince when you read such poor English. It would never occur to Snow to say, "Yes. This is he." *Wince* means to tense up from pain.

Snow: "Wow. He's the famous kid who teaches math at the university. Everyone knows about Fred Gauss. He's such a lovable guy."

Coalback: "Okay. Cut out the sweet talk. How much does he have in his bank account?"

Snow: "You know I'm not supposed to tell you this."*

Coalback: "Look Lester. I pay you good money each month for this information. What's the number?"

Snow: "Fred has $1,935.06 in his checking account."

Smiling, Coalback walked out and said, "Well, Fred Gauss, it looks like we have a deal."

Fred began to write a check:

FRED GAUSS
ROOM 314, MATH BUILDING
KITTENS UNIVERSITY

322

Date ____________

Pay to the order of C. C. Coalback Bicycles $ ________

____________________ dollars

Kittens Bank

"IT'S IN THE KITTY" ____________

"Don't forget," said Coalback, "There's sales tax and other stuff. The total comes to $1,935.06."

Fred thought *How lucky I am. If it had been a penny more, I wouldn't have been able to buy this bike.* He finished writing the check and handed it to Coalback.

Coalback grinned and said, "I'll get the bike shipped over to you by campus mail. It'll be there before you get back to your office."

---

* What they are doing is nefarious. [ni-FAIR-ee-us] This means very wicked.

For once, Coalback had said something that was true. The campus mail system at KITTENS University is very fast. There are mailboxes on every corner. And whenever anyone puts anything in the mailbox, a light on the top of the box lights up. A campus mailman comes and picks it up and rushes to deliver it.

The motto for the KITTENS University mail system is, "We're faster than email."

After Fred left the store, Coalback put the bicycle box into the mailbox. Then he headed back to his store and closed it. He closed it permanently. There was no more bicycle store.

Fred was whistling as he walked home. He was dreaming about all the places he would ride his bike. At first he thought that he would ride it half of the time to class and half of the time to the library.

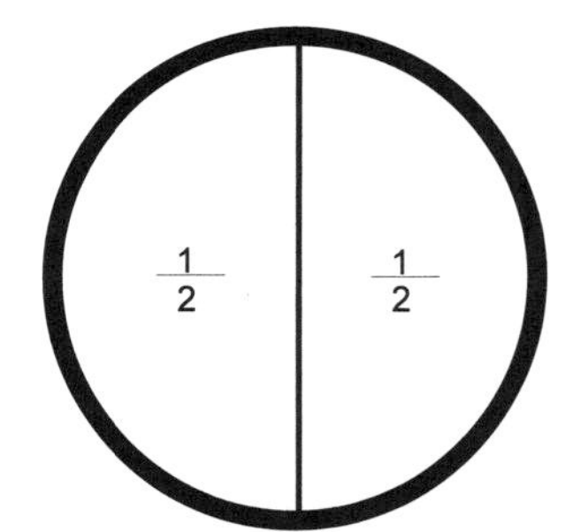

A perfectly balanced life of riding my bike. Two equal parts. Half the time riding to class and half the time riding to the library.

Oh! I almost forgot. I want to ride my bike in the park. And to Betty's house.

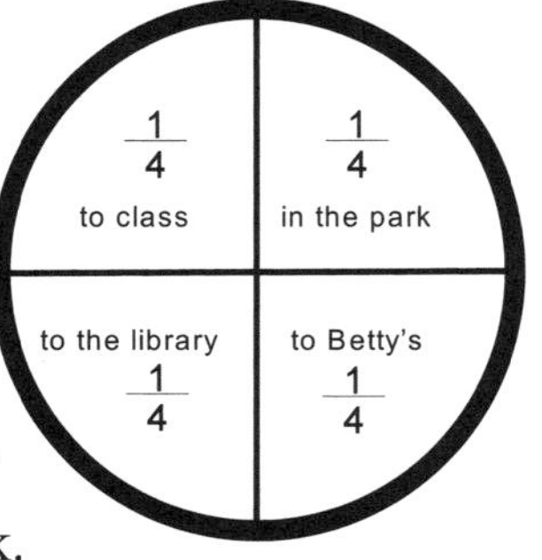

Four equal parts. Each part is one-fourth of the whole. One-fourth of the time Fred planned to ride in the park.

Your Turn to Play

1. Fred thought some more about where he wanted to ride his bike. He also wanted to ride to Alexander's house. Like Betty, Alexander is a student at KITTENS University. Fred has known them both for a long time. Fred would ride his bike to the pizza place called PieOne where they would often meet.

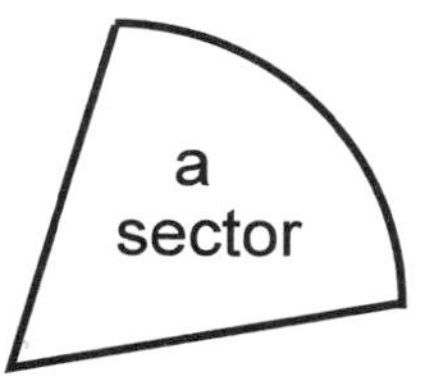

Draw a circle and divide it into 6 equal **sectors**.

2. Find the sum of one billion and four billion.
   Find the sum of means "add."
3. If Fred rode his bike for four hours, how many minutes would that be?
4. If Fred rode for one-sixth of those four hours in the park, how many *minutes* would he have ridden in the park?
5. What food comes in the shape of a sector?
6. Fred now has \$0.00 in his checking account. If he had three times that much, how much would that be?
7. Fred sleeps under his desk every night. He is 36" tall.

   Suppose that there are x inches from the floor to the underside of the desk.

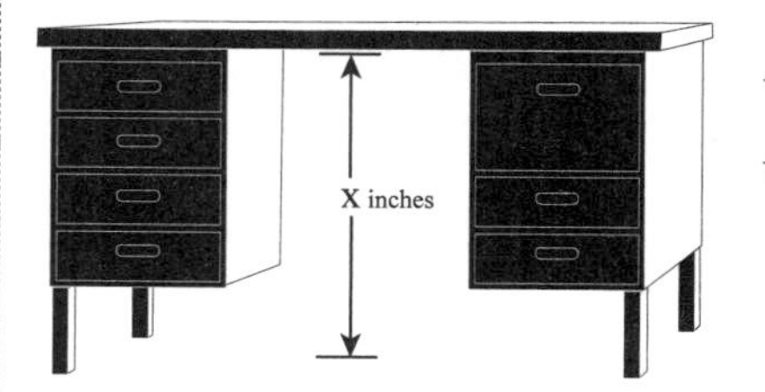

Which is better? x < 36" or x > 36"
We don't want Fred to hit his head when he stands up in the morning.

8. In symbols, *y is greater than or equal to 7* is written $y \geq 7$.
Guess how you would write *z is less than or equal to 55.*

Complete solutions are on page 170

Chapter Eight
Comparing Fractions

Fred stopped whistling and the notes floated off into the Kansas air. He realized that he wouldn't ride his bicycle  the same amount of time for each of the places he wanted to go.

On the KITTENS campus, there were two places he would ride to. He would ride to the classes he taught, and he would ride to the library.

Here is the notebook that Fred carries in his pocket:

TODAY

8:00 A.M.	BEGINNING ALGEBRA
9:00 A.M.	ADVANCED ALGEBRA
10:00 A.M.	GEOMETRY
11:00 A.M.	TRIG
12 M.*	CALCULUS
1:00 P.M.	TRIP TO THE LIBRARY

He taught five classes and went to the library only once.

The time he spent riding on campus was not:

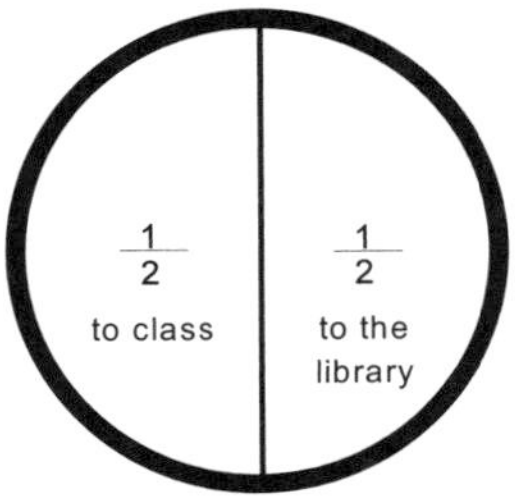

$\frac{1}{2}$ to class

$\frac{1}{2}$ to the library

* "M." stands for meridiem, which is noon. "A.M." is short for ante meridiem. The prefix *ante-* means happening before. So ante meridiem means happening before noon.

Instead, his riding on campus was five parts of going to classes and one part of going to the library.

One-sixth ($\frac{1}{6}$) of the time, he headed to the library.

Five-sixths ($\frac{5}{6}$) of the time, he was riding to classes.

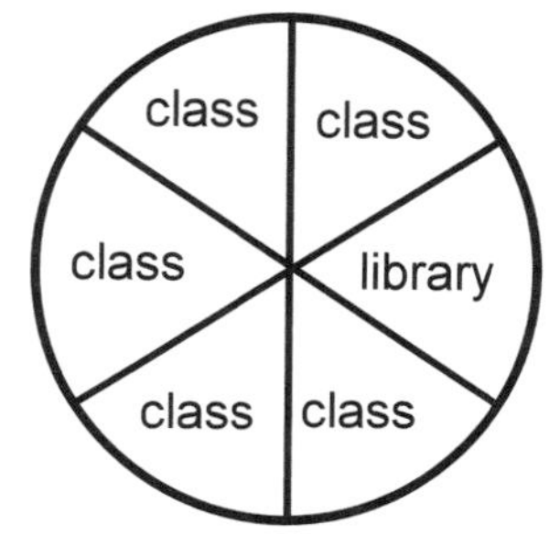

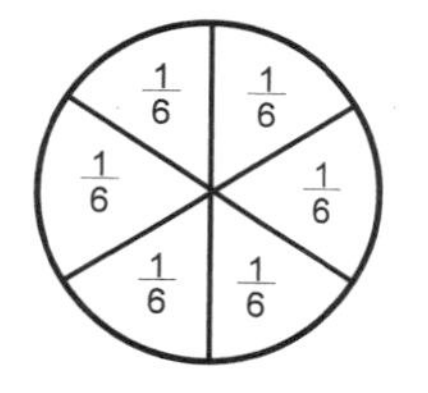

Six equal sectors.
Five of them ☞ to classes.
One of them ☞ to the library.

Do you like coloring? I do.

Here is what $\frac{5}{6}$ looks like:

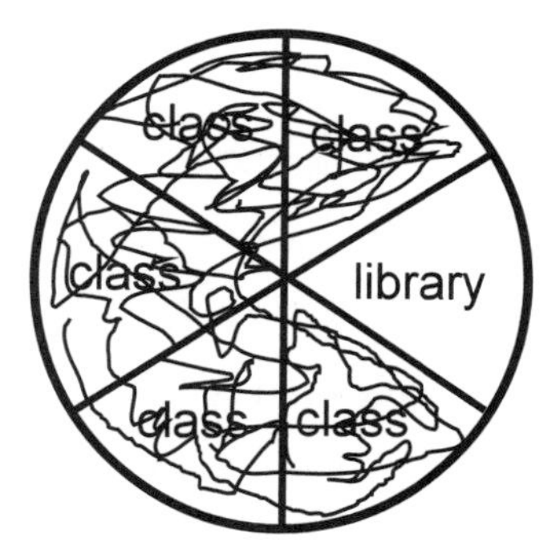

$\frac{5}{6}$ means

we have six equal pieces

and we take five of them.

Your Turn to Play

1. Draw a circle with four equal sectors. Color in $\frac{3}{4}$

2. We know that $\frac{1}{6} + \frac{1}{6} + \frac{1}{6} + \frac{1}{6} + \frac{1}{6} = \frac{5}{6}$

   What does $\frac{1}{4} + \frac{1}{4} + \frac{1}{4}$ equal?

3. Draw a circle and color in $\frac{1}{6}$ and
   draw another circle and color in $\frac{1}{4}$

   Which is true: $\frac{1}{6} < \frac{1}{4}$ or $\frac{1}{4} < \frac{1}{6}$?

4. Fred had \$1,935.06 in his checking account. He spent \$1,935.06 for his bike. How much did he have left?

5. If Betty had four thousand, twenty-six dollars in her checking account and she spent thirty-nine dollars, how much did she have left?

6. Which is true: $\frac{1}{10} < \frac{1}{3}$ or $\frac{1}{3} < \frac{1}{10}$?

7. Add 476 + 39 + 888.

8. Ten times ten is a hundred. $10 \times 10 = 100$

   Ten times ten times ten is a thousand. $10 \times 10 \times 10 = 1{,}000$.

   How many tens would you multiply together to get a million?

9. Suppose you have a pie cut into six equal pieces. Fred eats one piece. Betty eats one piece and Alexander eats two pieces. What fraction of the pie have they eaten?

$$\frac{1}{6} + \frac{1}{6} + \frac{2}{6} = ?$$

Complete solutions are on page 171

Fred walked back to the Math Building and climbed the stairs to the third floor. Outside the door of his office was his new bicycle. He opened the door and pushed the box inside. He said to himself I'll never have to walk to class again. I'll be taller riding on a bike. I'll get to places quicker. If it has a basket, I'll be able to carry more books home from the library each day.

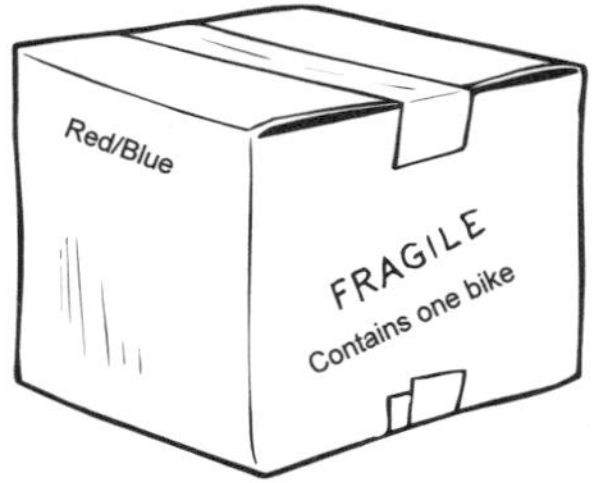

He reached into his desk drawer and got a pair of scissors so that he could cut through the tape on the box. Since Fred was only 5½ years old, the university gave him scissors with rounded ends so that he wouldn't hurt himself. The fact that he was a math professor at a university didn't make a difference. He was still only 5½ years old. The scissors with the rounded ends didn't work. They couldn't cut the tape.

rounded ends

What shall I do? I need something sharper. Something with a point on the end so I can cut the tape. Then he thought of the birthday gift that he had bought for Alexander. It was a knife that Alexander could use when he went camping. Since Alexander was going to be 21 years old, he was old enough to have a sharp knife with a point on the end. Fred took it out of a box in the bottom drawer of his desk and set it on his desk. It was a big knife—eighteen inches long.

18"

Fred thought to himself I'm 36" tall and it is 18" tall. Fred > knife. If the knife were a radius, I would be a diameter since 36" = 2 × 18". Fred was being silly.

He picked up the knife and decided to practice with it before he used it on the box.

He took an old comb out of the garbage and neatly cut it into six equal pieces.

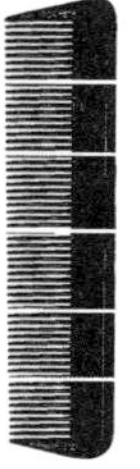

Your Turn to Play

1. Look at the comb on the right. Each piece is what part of the whole comb? Find the number that replaces the question mark. $\frac{1}{?}$

2. How many pieces of the comb would you need to make one-half of the whole comb?

3. If there were six pencils, and Betty and Alexander each take an equal share of them, what fraction of the pencils would each person get?

4. If we take those six pencils and divide them equally among Fred, Betty, and Alexander, what fraction of the pencils would each get?

5. If there were 30 paper clips and they were divided equally among Fred, Betty, and Alexander, what fraction of the paper clips would each get?

6. In the above problems, we found that

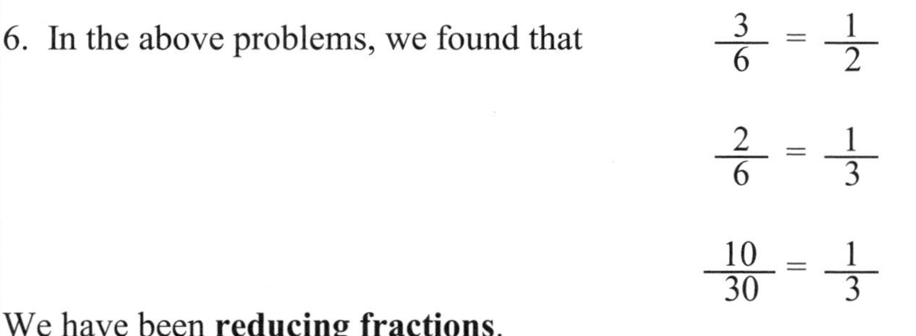

$$\frac{3}{6} = \frac{1}{2}$$

$$\frac{2}{6} = \frac{1}{3}$$

$$\frac{10}{30} = \frac{1}{3}$$

We have been **reducing fractions**.

Here's an easy way to reduce a fraction (without having to draw a lot of pictures): *You can divide the top and bottom of a fraction by the same number.*

If I start with $\frac{2}{8}$ and I divide top and bottom by 2, I get $\frac{1}{4}$

If I start with $\frac{4}{6}$ and I divide top and bottom by 2, I get $\frac{2}{3}$

Your turn: reduce $\frac{7}{21}$ (Divide top and bottom by 7.)

Complete solutions are on page 172

Chapter Ten

Add and Reduce

There was a knock on the office door. It was Betty. "It's me. May I come in?" Fred pushed the six pieces of the comb into his waste basket. He didn't want her to see that he was playing with a big knife. He was excited. He wanted to show her his new bike.

Five-year-olds sometimes make mistakes. Even when you are five-and-a-half years old you make mistakes. When you get to be six years old, then you don't make any mistakes anymore.✶

Quiz time! Which of these is a mistake?

A) Betty knocked on the door.

B) Betty asked, "May I come in?"

C) Fred was excited to see Betty.

D) Fred ran to the door carrying this big old knife.

He dropped the knife. It sort of slipped out of his hands.

The Good News:

✧ The knife didn't break.

✧ He didn't cut Betty.

The Medium News:

✧ The knife fell to the floor. (That's the way gravity works.)

The Bad News:

✧ The knife headed straight towards Fred's foot.

More Good News:

✧ Fred was wearing shoes.

More Bad News:

✧ The knife was big and heavy.

✶ That is a big fat untruth. Just because something is written in a book doesn't make it true.

Just before the knife fell, Fred called out to Betty, "Come on in."

Betty came in and saw Fred just standing there. Fred wasn't moving. Fred couldn't move. Fred was nailed to the floor.

Fred was going to say, "Hi Betty. I want you to look at the new bike I bought." But he didn't say that.

He was thinking of saying, "Hi Betty. I just dropped a knife on my foot." But he didn't say that.

His mouth opened, and he wanted to say, "Help!" But nothing came out of his mouth.

Instead, he fainted.

As we noted at the beginning of this book, when Fred first came to KITTENS University he was only nine months old, and Betty would often carry him to class. It was time to carry Fred again.

She picked him up (after unsticking the knife from the floor). He wasn't that heavy—only 37 pounds. He was a very light five-and-a-half-year-old boy. And Betty didn't have far to carry him. The hospital was only a block from the Math Building.

Now, in the movies, when someone gets hit with an arrow or stabbed with a knife, they pull it out. But that makes it bleed a lot more. Sometimes it's better just to leave it in until you get to the doctor. Since this is real life—and not the movies—Betty left the knife in his foot.

As she carried him down the stairs in the Math Building, Betty was surprised that she was starting to get tired already. A thirty-seven-pound load wasn't much for her to carry. She had forgotten about the thirteen-pound knife.

Your Turn to Play

1. How much weight was Betty carrying?
2. What fraction of the total weight was Fred?
3. What fraction of the fifty pounds was the knife?
4. If you add Fred's fraction of the total weight ($\frac{37}{50}$) and the knife's fraction of the total weight ($\frac{13}{50}$), what do you get?
5. Reduce $\frac{15}{20}$ (Divide top and bottom by 5.)
6. Add and reduce your answers: $\frac{1}{8} + \frac{1}{8} = ?$

   $\frac{2}{9} + \frac{1}{9} = ?$
7. A mile is equal to eight furlongs. It's 93 million miles from the earth to the sun. How long is that in furlongs?

Complete solutions are on page 173

The Bridge

from Chapters 1–10 to Chapter 11

first try

Goal: Get 9 or more right and you cross the bridge.

1. Double 1 and you get 2. Double again, and you get 4. Keep doubling until you get a number larger than a thousand. How many times did you have to double?

2. Find the sum of 783 and 697.

3. Draw a circle and divide it into seven roughly equal sectors.

4. Fred weighs 37 pounds. How many ounces is that? (There are 16 ounces in a pound.)

5. If it takes Betty 7 minutes to get Fred to the hospital, how many seconds is that?

6. Seven minutes is what fraction of an hour?

7. A sign in front of some roller coaster rides reads, "You should be at least 42 inches tall to go on this ride." Let x stand for your height. Which of these is the best way to put the sign into symbols?

$x < 42$ $\quad x \leq 42$ $\quad x = 42$ $\quad x > 42$ $\quad x \geq 42$

8. Which is larger: $\frac{1}{8}$ or $\frac{1}{6}$?

9. Add and reduce your answer: $\frac{1}{10} + \frac{4}{10}$

10. If you have one dollar less than a billion dollars, how much would you have? Give your answer in words, not numbers.

If you got nine or ten correct, congratulations! You are ready to go on to Chapter 11, which begins on p. 57.

Most people don't cross the bridge on the first try.

To earn the right to a second try, first correct all the errors you have made.

The Bridge

from Chapters 1–10 to Chapter 11

second try

1. Name the smallest ordinal number.
2. Zero times one billion equals what?
3. Draw a picture of a sector.
4. Mt. Everest is 29,028' tall. If Fred were 53' tall [subjunctive mood], how much taller would Mt. Everest be than Fred?

5. Ten minutes is what fraction of an hour? Reduce your answer as much as you can.
6. 121" is how many feet? Express your answer in feet and inches.
7. Find the sum of 42,973 and 558.
8. If I could save seven dollars every minute, how much could I save in a week?
9. Add and reduce your answer: $\frac{4}{9} + \frac{2}{9}$
10. Reduce as far as possible $\frac{36}{144}$

Did you know that some people always do all five bridges, even though they get nine or ten right on the first bridge? They do them all just to get in some extra practice.

The Bridge

from Chapters 1–10 to Chapter 11

third try

---

1. Eight minutes is what fraction of an hour? Reduce your answer as much as you can.

2. Which is larger: $\frac{1}{8}$ or $\frac{3}{8}$?

3. Persons are considered a millionaire if their wealth is at least a million dollars. (It doesn't all have to be in cash. Part of it could be in stocks, bonds, real estate, gold, or art.) If your wealth is equal to y and you are a millionaire, which of these must be true?

$y < \$1,000,000$ $y = \$1,000,000$ $y > \$1,000,000$ $y \geq \$1,000,000$

4. Ten times ten is a hundred. $10 \times 10 = 100$. How many tens would you multiply together to get a billion?

5. C. C. Coalback is not a very honest person. Suppose he had to drink seven tablespoons of habanero* sauce for every lie he has told. So far in his life, he has told 8,379 lies. How many tablespoons of habanero will he have to drink?

6. Reduce as far as possible $\frac{25}{65}$

7. Find the sum of two billion and four hundred billion. Write your answer as a number.

8. If I could save eight dollars every hour, how much could I save in a week?

9. Mt. Everest is 29,028' tall. Annapurna is 26,503' tall. If you stacked Annapurna on top of Mt. Everest, how tall would this super mountain be? (We could call it Mt. Fred.)

10. 777" is how many feet? Express your answer in feet and inches.

---

* habanero [ha-beh-NARE-oh] This is a small pepper with a really big "bite"—a very hot pepper. One drop is too much for many people.

The Bridge

from Chapters 1–10 to Chapter 11

fourth try

1. The knife in Fred's foot weighs thirteen pounds. How many ounces is that? (There are 16 ounces in a pound.)

2. Which is larger: $\frac{1}{8}$ or $\frac{1}{10}$?

3. *Drip!* is the onomatopoetic word describing Fred as Betty carried him down the stairs. If Fred were losing one drop of blood every four seconds, how many drops of blood would he lose in four minutes?

4. Reduce as far as possible $\frac{120}{360}$

5. Double 3 and you get 6. Keep doubling until you get a number greater than two thousand.

6. Add and reduce your answer: $\frac{1}{12} + \frac{5}{12}$

7. If your wealth is equal to x dollars, and if you are a billionaire (a person who has at least a billion dollars), which of the following must be true?
$\$1,000,000,000 < x$
$\$1,000,000,000 \leq x$
$\$1,000,000,000 = x$
$\$1,000,000,000 > x$
$\$1,000,000,000 \geq x$

8. The largest desert in the world is the Sahara. It's located in the northern part of Africa. It is three million, five hundred thousand square miles. The second largest desert in the world is the Great Australian. It is 1,480,000 square miles. Is the Great Australian less than half the size of the Sahara?

9. It is seven furlongs from Fred's office to the hospital. How many feet is that? (A furlong is 220 yards.)

10. Find the sum of fifty-five million and twenty-six billion. Write your answer as a number.

The Bridge
from Chapters 1–10 to Chapter 11

fifth try

1. If it takes Betty 8 minutes to get Fred to the hospital, how many seconds is that?

2. Find the sum of one hundred forty-six and six hundred seventy-seven. Write your answer as a number (not as words).

3. Ten times ten is a hundred. $10 \times 10 = 100$. How many tens would you have to multiply together to get ten billion?

4. Add and reduce your answer: $\frac{3}{14} + \frac{4}{14}$

5. Betty weighs 125 pounds. How many ounces is that? (There are 16 ounces in a pound.)

6. Convert 78" into feet and inches.

7. An art question: Draw a circle and divide it into five roughly equal sectors.

8. $0 \times 323{,}500{,}899{,}222{,}714 = ?$

9. Thirty minutes is what fraction of an hour? Reduce your answer as much as you can.

10. If Fred could save $493 each year, how much could he save in a century? (A century is a hundred years.)

If you got nine or ten right, turn to the next page to start Chapter 11. If not, then please correct your errors and turn back four pages to "first try."

Chapter Eleven

Subtracting Fractions with the Same Denominators

Betty was breathing hard by the time she got to the bottom of the stairs. Fred and the knife weighed 50 pounds and she weighed 125 pounds. That was 175 pounds on her feet. Then she saw Alexander.

Alexander ran over to Betty. "What's going on?" Alexander asked.

"Please help me get him to the hospital," Betty said. She transferred Fred into Alexander's arms.

Fred's eyes were starting to open, but he was still a little dizzy. He mumbled something about this being Alexander's birthday present, but he didn't have a chance to wrap it. (He was talking about the knife.)

In a minute, they got to the hospital and went inside. The nurse at the desk said, "Hi. What seems to be wrong?"

This was a very silly question. Alexander was carrying a 37-pound boy with a 13-pound knife in his foot. And Fred's foot was dripping blood on the hospital carpet.

"We'd like to see a doctor," Alexander answered.

"Of course you would," the nurse said. "That's why people come to the hospital. They don't come here to buy doughnuts. Please fill out this form."

Betty took the form, sat down, and filled it out.

New Patient Form

Name: Fred Gauss

Address: Room 314, KITTENS

Age, Weight & Height: 5, 37, 36

What hurts? His foot.

What should we do about it? Fix it.

Betty handed the form back to the nurse.

"It will only be a minute," the nurse said.

Alexander sat down with Fred on his lap. He said to Fred, "It will be okay. We will be with you during this whole thing. Just relax." Fred shut his eyes and put his head against Alexander's chest. Every four seconds you could hear, *Drip! . . . Drip! . . . Drip!*

Betty paced back and forth in the waiting room. Fred dripped. After fifteen minutes, Betty went back to the nurse at the desk and asked how long it would be until Fred could see a doctor.

"It will only be a minute."

Another twenty minutes passed. *Drip! . . . Drip! . . . Drip!* When Betty asked the nurse again, the nurse came out from behind the desk and walked over to look at Fred.

"He's bleeding," the nurse said. "He should see a doctor right away."

The nurse took a meter out of a bag marked "Auto Repair" and hooked it up to Fred's arm.*

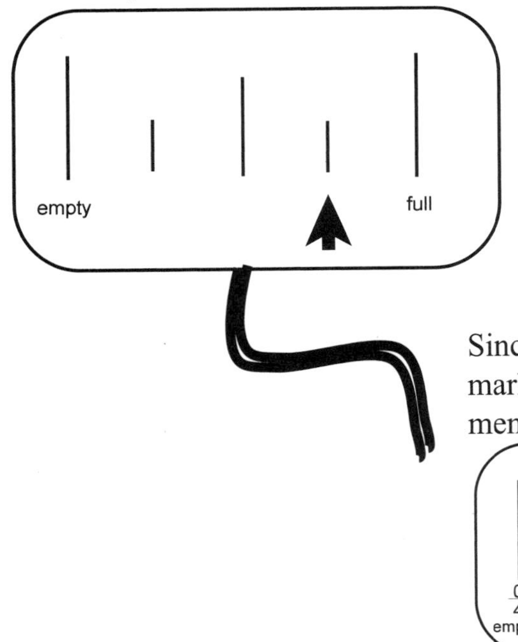

"Just as I thought," the nurse said. "He's down a pint."

Alexander looked at the gauge and shouted, "He's lost a quarter of all his blood!"

Since there are five equally spaced marks on the gauge, Alexander had mentally inserted the fractions:

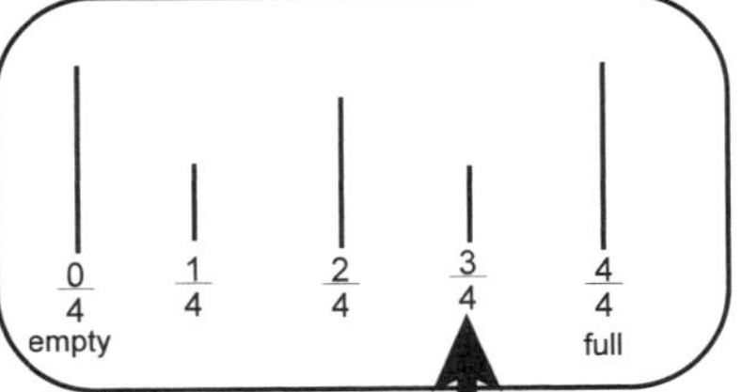

✶ The nurse used to work at Auto Repair—where they say they can change the oil in your car in less than 26 minutes. Their sign read: `Oil Change < 26 Minutes`.

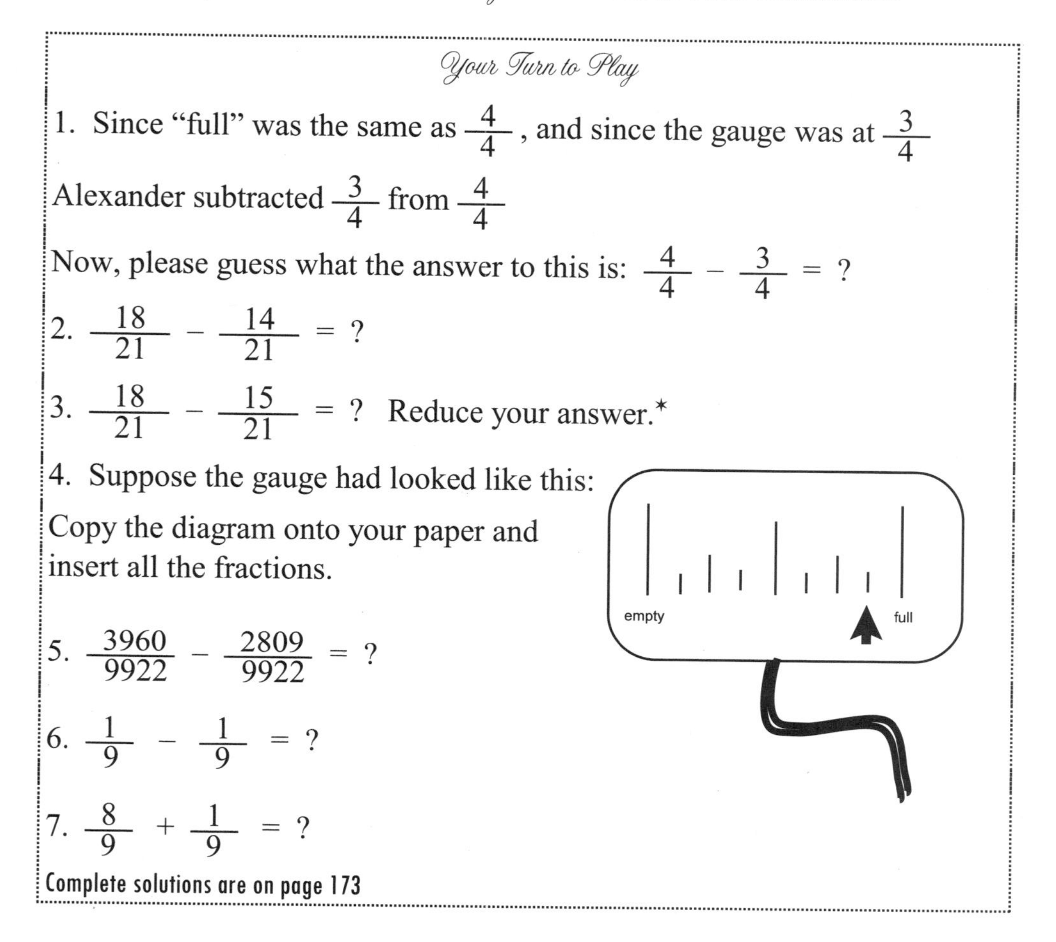

Your Turn to Play

1. Since "full" was the same as $\frac{4}{4}$, and since the gauge was at $\frac{3}{4}$

Alexander subtracted $\frac{3}{4}$ from $\frac{4}{4}$

Now, please guess what the answer to this is: $\frac{4}{4} - \frac{3}{4} = ?$

2. $\frac{18}{21} - \frac{14}{21} = ?$

3. $\frac{18}{21} - \frac{15}{21} = ?$ Reduce your answer.*

4. Suppose the gauge had looked like this:
Copy the diagram onto your paper and insert all the fractions.

5. $\frac{3960}{9922} - \frac{2809}{9922} = ?$

6. $\frac{1}{9} - \frac{1}{9} = ?$

7. $\frac{8}{9} + \frac{1}{9} = ?$

Complete solutions are on page 173

✶ It is now time for a General Rule:

General Rule

Please reduce fractions in your answer as much as possible.

Chapter Twelve
Common Denominators

Another twenty minutes passed. Fred, Betty, and Alexander had been waiting for fifty-five minutes. Betty was getting angry, and Fred was turning as white as a sheet of paper.

Alexander looked at the meter that was hooked up to Fred's arm. It didn't look good.

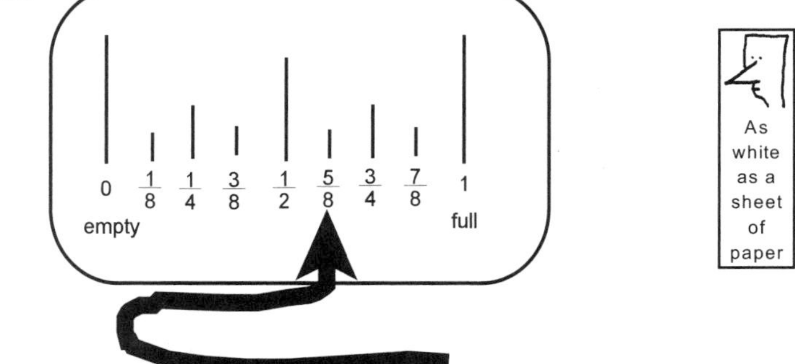

$\frac{5}{8} < \frac{3}{4}$ You can tell that by just looking at the meter. The numbers on the left are smaller than the ones on the right.

But just looking at $\frac{5}{8}$ and $\frac{3}{4}$ it is hard to tell which is smaller. Before we reduced the fractions, it was easier.

It's easier to see that $\frac{5}{8} < \frac{6}{8}$

(Recall that $\frac{6}{8}$ and $\frac{3}{4}$ are equal.)

To compare two fractions—to see which one is smaller—you need to have their bottoms alike.*

To compare two fractions, they need to have a common (the same) denominator.

---

* It's probably time to use the grown-up words. Instead of $\frac{\text{top}}{\text{bottom}}$ most mathematicians say $\frac{\textbf{numerator}}{\textbf{denominator}}$

They would say that to compare two fractions—to see which one is smaller—you need to have equal denominators.

It's okay if you still want to talk about the bottoms of fractions. I'll know what you mean. When my daughter was just a baby, her denominator was sometimes wet.

Earlier, we reduced $\frac{6}{8}$ to $\frac{3}{4}$ by dividing numerator and denominator by 2.

Now we go in the opposite direction and change $\frac{3}{4}$ to $\frac{6}{8}$ by multiplying numerator and denominator by 2.*

Your Turn to Play

1. Which is smaller: $\frac{1}{4}$ or $\frac{5}{12}$? (Change the $\frac{1}{4}$ to a fraction that has a denominator equal to 12.)

2. Compare $\frac{2}{3}$ and $\frac{7}{12}$

3. Now, here's a brain buster. You haven't seen a problem quite like this one. In the previous two problems you changed *one* of the fractions so that it would have the same denominator as the other fraction.

Compare $\frac{2}{5}$ and $\frac{3}{8}$

4. Now with the meter reading $\frac{5}{8}$ what fraction of all of his blood has Fred lost?

(When the meter read $\frac{3}{4}$ he had lost one-quarter of all his blood.)

5. Betty's resting pulse rate is 65 beats/minute. It was now at 84 beats/minute. How much faster is it now?

6. They have waited for fifty-five minutes. How much longer until they have waited for a whole hour?

Complete solutions are on page 174

* You can take any fraction and multiply top and bottom (oops, numerator and denominator) by the same number. The answer you get is equal to the original fraction. Take any old fraction such as $\frac{30}{41}$ and multiply numerator and denominator by 5. You get $\frac{150}{205}$ and this is equal to the original $\frac{30}{41}$

Chapter Thirteen
Roman Numerals

Finally, the nurse led them down the long hall to an examination room. Betty walked quickly trying to hurry things up. Alexander carried Fred. Fred seemed to be on the edge between sleeping and waking. He said something like, "All the golly big chocolate frog."

"The doctor will be with you in a minute," the nurse said as she closed the door on the examination room. The doors each had a letter on them: A, B, C, D. . . . Their door was marked with a W.

The dripping had slowed down. It was now only once every thirty seconds that you could hear *Drip!*

There was a knock on the door, and the doctor entered. "And what seems to be wrong?" he asked.

Betty pointed toward Fred. Instantly, the doctor noticed the 13-pound knife in Fred's foot, and, instantly, he made his diagnosis*: "He's bleeding." (It had taken the nurse 35 minutes to figure that out. Doctors have more years of medical training than nurses.) "This is most unusual," he continued. He scratched his head, sat down, and pulled out Prof. Eldwood's *Guide to Modern Medicine*, 1849. The doctor didn't know whether to look up *foot* or *knife* or *bleeding.* In his wisdom he decided to look up *bleeding.*

Guide to Modern Medicine

billing: Don't forget to do this.

bladder: The kidneys are connected to the ureters which are connected to the bladder which is connected to the urethra which is connected to "the exterior." Don't forget to empty.

bleeding: Stop it.

blood: Red stuff in veins.

blud: See *blood.*

p. 63

"It appears," the doctor began, "that the patient has a bleeding problem. It is important that we . . . stop it."

He got up and walked over to Fred. "I guess we'll have to get this out of the way so that I can see what's going on." He pulled the knife out of Fred's foot, wiped it off, and put it in his briefcase. "No use throwing a good knife away," he said to himself.

---

* diagnosis [die-ag-KNOW-sis] Looking at something and figuring out what's wrong.

Blood squirted from Fred's foot. It gushed. It splashed. It flowed. It ran. Doctors are trained in medical school never to say "Oops!"—especially in front of patients. It makes the patients nervous. Instead, he said, "I'll be back in a minute" and left the room.

Inside the room Betty sprang into action. She pulled off Fred's shoe, took her handkerchief and wrapped it around Fred's foot to stop the bleeding. Fred has a small foot, and the handkerchief went around his foot eight times. In a minute the bleeding had stopped.

Outside the room the doctor wiped his forehead. He saw the janitor in the hall who was mopping up the drops of blood that Fred had lost as Alexander had carried him down the hall to the examination room.

"Looks like you got a bleeding problem," the janitor said to the doctor.

"Yeah. He's lost of a lot of blood," the doctor replied. "I wonder what we should do."

The janitor opened the door marked W and peeked inside. He turned back to the doctor and said, "The meter on his arm says he's about half full. It's time you stick a Two Plus Two into him and give him some blood."

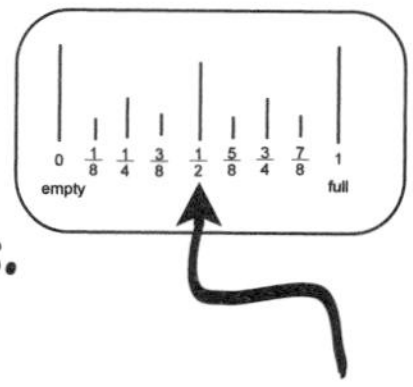

Wait! Stop! I, your reader, have a lot of questions.

Okay. Ask your questions.

First of all, if you lose half the blood in your body, wouldn't you be dead?

Fred is a pretty tough little kid.

Second, what is a "Two Plus Two"? I've never heard of that before.

You know what 2 + 2 is equal to?

Sure. Everyone does. It equals 4. But that still doesn't make sense. What does it mean to "stick a 4 into him and give him some blood?"

And what is 4 in Roman numerals?

It's IV.

An IV is explained in the footnote,[*] and Roman numerals are explained on the next page.

---

[*] IV stands for intravenous ("into a vein"). Roughly speaking, it means sticking a needle into your arm to give you some medicine, blood, or other stuff.

Before I explain Roman numerals, I have to tell you about Arabic numerals. For the last 900 years (since the 12th century) we have been using Arabic numerals. They are 0, 1, 2, 3, 4, 5, 6, 7, 8, and 9.

Wait! I haven't been alive for 900 years.

When I said "we," I was including your parents.

Okay.

With these Arabic numerals you can make numbers like a billion (1,000,000,000).

Numerals . . . numbers—what's the difference?

Okay. Take this little test. Tell me which is bigger:

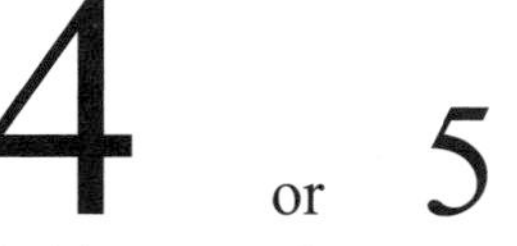

The numeral 4 is bigger. The **numeral** is the symbol. It's what you write down.

The number 5 is bigger. The **number** is "how many."

Before the 14th century Roman numerals were commonly used.

Wait! I hate to keep interrupting, but your math seems to be a bit funny. At the top of this page you wrote that Arabic numerals have been used since the 12th century, and now you just wrote that Roman numerals were used up to the 14th century. Which is it?

Both. There was a period when both were commonly used. Back in the "old days" it took lots of time to make changes. Now things happen a bit more quickly. Email, for example, didn't take two centuries to become popular.

I have one last question before you start talking about Roman numerals. Why do I have to learn about them if they are no longer used? I hate wasting my brain on stuff I'll never use.

There is a bunch of places where Roman numerals are still used today. For example,

- Super Bowl XXIV
- Queen Elizabeth II
- World War II
- page xiv (The pages at the very beginning of some books are numbered using lowercase Roman numerals.)

✳ Films often use Roman numerals to make it hard to figure out when they were made. This is especially useful when they are broadcast on television. All you see briefly at the end of the credits is ©MCMXCI, and you wonder, "Was this before I was born?"

Roman numerals use seven letters: I, V, X, L, C, D, and M. Their values are I = 1, V = 5, X = 10, L = 50, C = 100, D = 500, and M =1000. With these you can count up to about 3000 or so. Let's begin:

I = 1

II = 2

III = 3 ***Hey! So far, this isn't hard.***

IV = 4 (That's one less than five.)

V = 5

VI = 6 (That's one more than five.)

VII = 7

VIII = 8

IX = 9 (That's one less than ten.)

X = 10

I have a question. Why don't they use IIII for 4 and VIIII for 9?

I would guess that it's because IIII is harder to read than IV. The only place I've seen IIII is on some clock faces.

Your Turn to Play

1. Write in Roman numerals the numbers from 11 to 20.

2. Write in Roman numerals the numbers 60, 90, and 400.

3. The "number of the beast" as mentioned in the Book of Revelation is 666. What is peculiar is that when 666 is written in Roman numerals, it uses all the basic symbols except M. Express 666 in Roman numerals.

4. For advanced students of Roman numerals, there is the rule for which letters can be used in the "less than" notation. You write IX for 9, but you can't write IC for 99. *The letter I can go on the left of only V or X.* To write 99, you write XCIX. The "XC" is the 90 and the "IX" is the nine.

The rule continues: *The letter X can go on the left of only L or C. The letter C can go on the left of only D or M.*

So when I write IC, I mean ice cream.

Here is the ultimate challenge: Using the rules, write 1999.

5. The real agony comes when you try and do arithmetic with Roman numerals. If someone asked me to divide DXLVI by XIV, I wouldn't start by writing $\text{XIV}\,\overline{)\,\text{DXLVI}}$. It would take ten weeks to do it that way. Here's an essay question: If someone locked you in a room with a ton of paper and a pencil and told you to divide DXLVI by XIV, how would you go about doing it? Don't do it. Just describe how you would go about doing it.

6. Now do it. Divide DXLVI by XIV and give your answer in Roman numerals.

Complete solutions are on page 175

Chapter Fourteen
Adding Fractions

The janitor's suggestion was a good one. The doctor passed that suggestion along to the nurse at the front desk. She came and stuck a "Two Plus Two" into Fred's arm and gave him a refill.

Alexander was delighted to see Fred's eyes open. Fred looked at him and said, "Sky rocked duh life."

As more blood flowed into Fred, he said, "Dry docked lay life."

Then, "My dropped the wife."

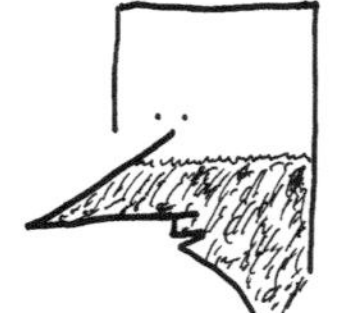

Betty looked at Fred and told the nurse, "It looks like he could use another pint. He's not quite full yet."

After Fred received another pint, he said, "I dropped the knife." At last, he was starting to make sense again. He said to Alexander, "I was trying to open the bicycle box with the knife that I bought you for a birthday present since the scissors with the round end that the university gave me didn't cut the tape that held the box together and so I got the knife out of the box in the bottom drawer and was going to use it on the tape after I practiced on a comb, which I cut into six pieces, when Betty came to visit me, and I swept the six pieces of comb into the waste basket so that she wouldn't see that I was playing with the knife, and then I raced to the door to let Betty in, and in doing so, I lost control of the knife, and it slipped, and I don't remember what happened after that."*

After all that he said a very short sentence: "My foot hurts."

Betty explained to him what had happened in the last hour-and-a-half since the accident.

"Hey, it's almost noon," Alexander said as he looked at his watch. "Let's all go and get some lunch."

"Good idea!" Betty said, and the three of them walked out of the hospital. Actually, two of them walked, and one was still being carried.

They headed toward their favorite place to eat.

---

* That is probably the longest sentence ever written in a math book. If you are curious, it is 141 words long.

PieOne had been a favorite place for many KITTENS University students over the years. Birthday parties were always celebrated at PieOne. Betty always liked to bring Fred here to eat. She was worried about how poorly he ate.

Living in room 314 in the Math Building and having no mother or father to look after him, Fred had gotten into some pretty bad eating habits. Much of the time he would buy his food from the machines that were down the hall from his office.

In the last half year Fred hadn't grown an inch. He was still 36 inches tall. There aren't many five-and-a-half-year-old boys who weigh only 37 pounds.

They entered PieOne. It was darker than most other pizza places. Plastic grape leaves were taped to the walls. Pictures of famous Italians, such as Chico Marx, were placed amid the grape leaves. It was lunchtime, and the place was filled with students. Betty, Alexander, and Fred found an empty table. Betty and Alexander put their coats on a chair, and Alexander put Fred on top of the coats. Fred was embarrassed when they used to ask for a booster chair for him, but he was too short and really needed one. The coats were the best way to make Fred tall enough.

Fred suddenly turned a little pink and said, "I can't eat here today. I just realized that I'm out of money."

"It's okay," Betty answered. "I took you off to the hospital before you had a chance to grab your keys and wallet and stuff. You can pay us back when we get back to your office."

A Quick Course in Adding Fractions

To add $\frac{2}{5}$ and $\frac{3}{8}$.

Step One: Make the bottoms alike.

Step Two: Add the tops and copy the bottom.

40 would be a good new bottom. Both 5 and 8 divide into 40 evenly.

$\frac{2}{5}$ becomes $\frac{16}{40}$ (multiplying top and bottom by 8)

$\frac{3}{8}$ becomes $\frac{15}{40}$ (multiplying top and bottom by 5)

Then add the tops and copy the bottom $\frac{16}{40} + \frac{15}{40} = \frac{31}{40}$ Done!

"But I can't pay you when we get back to my office," Fred exclaimed. "I'm broke. I spent all my money on my new bike."

"I don't get it," Alexander said. "I thought you had about two thousand dollars in your checking account."

"I did," Fred answered. "But the bike cost \$1,935.06, which is all the money I had."

Betty and Alexander looked at each other. They knew that bicycles for five-and-a-half-year-olds didn't cost *that* much. "How are you going to eat for the rest of this month?" Betty asked.

Fred shrugged his shoulders. Then he remembered the ad he had seen in the KITTEN Caboodle newspaper. He hopped off his chair and said, "Ouch." He had forgotten his hurt foot. He walked over to Stanthony and said, "I need a job. I can be a pizza cook. I can serve the pizzas. I can be a cashier. Whatever you want. I'm out of money. I spent it all on a new bike."

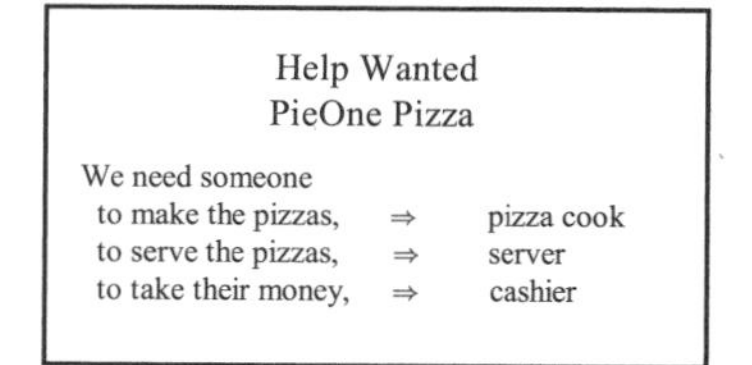
Help Wanted
PieOne Pizza

We need someone
to make the pizzas, ⇒ pizza cook
to serve the pizzas, ⇒ server
to take their money, ⇒ cashier

Stanthony looked at him. He had known Fred for years—ever since Betty had carried him into PieOne when he was only nine months old. "Well, my penniless, prodigal professor,* I'm sure we can find some work for you." He took Fred back to the kitchen. He tied an apron around Fred, but it was much too long. A dish towel worked just fine. Fred's arms were trapped under the towel. After a couple of minutes, he got them free.

"Okay. You are hired. Clean up this place and then start making pizzas."

Your Turn to Play

1. The kitchen was a mess. Fred started with the spice rack. There were two bottles of pepper. One had $\frac{1}{5}$ of a tablespoon of pepper and the other had $\frac{3}{5}$ of a tablespoon. He combined them into one bottle and threw the other bottle in the garbage. How much pepper did he now have?

2. He did the same with two bottles of basil: $\frac{1}{4} + \frac{1}{8} = ?$

3. He combined three bottles of oregano: $\frac{1}{3} + \frac{1}{4} + \frac{1}{6} = ?$

4. The garbage was starting to fill up with empty spice bottles as Fred cleaned up the kitchen. He looked at the $\frac{4}{5}$ of a tablespoon of pepper that he had (from problem 1) and the $\frac{3}{4}$ of a tablespoon of oregano that he had (from problem 3). Did he have more pepper or more oregano?

5. One bottle of ketchup was labeled, "Use before MMIV." Translate that into Arabic numerals.

6. So far, Fred has worked for ten minutes. What fraction of an hour is that?

7. Stanthony said that next month PieOne would be serving its one-millionth pizza. Is that a cardinal or an ordinal number?

Complete solutions are on page 176

✶ *P*enniless, *p*rodigal *p*rofessor is an example of alliteration—words that begin with the same sound. Used in poetry, speeches, and advertising, alliteration helps the listener remember what you said. Stanthony likes to use alliteration in his everyday talk, even though most of what he says isn't worth remembering.

A *prodigal* person is a wastrel, a spendthrift, one who spends way too much money.

Fred knew that in about a third of an hour* he would have all the spice bottles cleaned up. He was starting to think of himself as a chef. "Chef Fred" sounds nice he thought to himself. In addition to being a professor of mathematics, the university could let me be a professor of cooking.

Stanthony came into the kitchen. "Hey, my pensive partner in pizza production,** please put this platter of pepperoni pizzas on the table in the corner."

Fred picked up the platter and called to Stanthony, "Roger. Right! Ready to go." Fred was trying to do some alliteration, but wasn't very good at it.

He took the three pizzas out to the eight hungry students at the table in the corner.

"Professor Fred!" one of the students exclaimed. "What are you doing here serving pizzas?"

Fred blushed a little. "Just trying to earn a little extra money."

The students knew what Fred was talking about. None of them had a lot of money. Most college textbooks were so expensive. Luckily,

---

* one-third of an hour = ⅓ × 60 minutes = $\begin{array}{r}20\\ 3\overline{)60}\end{array}$ = twenty minutes

** More alliteration by Stanthony. *Pensive* = dreamily thinking

PieOne was a cheap place to eat. Three large pizzas cost $8, so each student paid a dollar. But the hard part was dividing up the pizzas equally. With three pizzas and eight students, how much should each person receive? Since they had all had Fred as their math teacher, they asked him.

"Okay. Tell us how much pizza should each of us get?"

That was an easy question for Fred. "Each person should get $\frac{3}{8}$ of a pizza."

"How did you get that?" they asked.

Fred had to switch into math teacher mode. He took off his chef's hat and untied the dish towel.

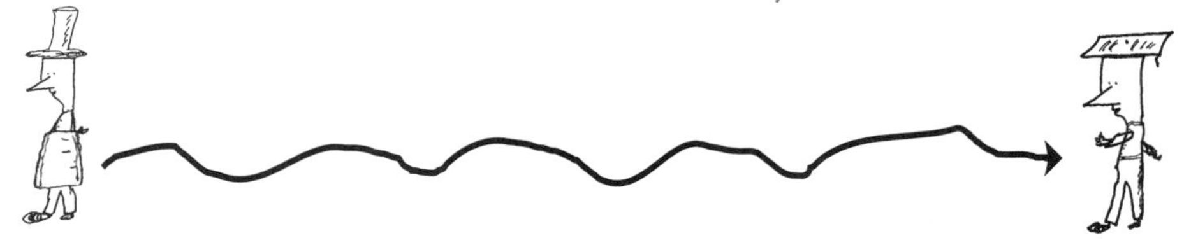

"Suppose we had one pizza and five students. How much would each get?" Fred asked.

"Easy. Each would get $\frac{1}{5}$" they responded.

"That's one pizza divided five ways," Fred pointed out. "If you divide a circle into five equal parts, each part is one-fifth."

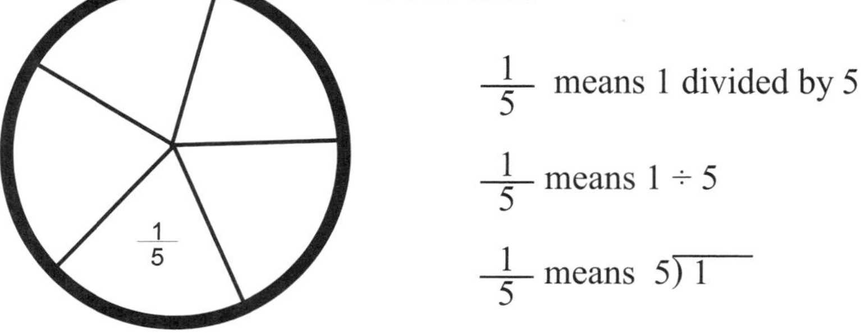

Fred continued, "So when you want to divide 3 pizzas into 8 equal parts, each part is $\frac{3}{8}$ Nothing could be simpler. If I had 7 pounds of sugar and I wanted to divide it equally among 43 kids, each kid would get $\frac{7}{43}$ pounds of sugar."

Your Turn to Play

1. If Joe and Darlene were on a pizza date, and Joe ate $\frac{3}{5}$ of the pizza that they ordered, what part of the pizza would be left for Darlene?

2. If Betty, Alexander, Joe, and Darlene shared three pizzas equally, how much would each person get?

3. Stanthony liked to hold contests at PieOne. One week the first prize was a tapir. Joe and Darlene had entered the contest together, and the tapir was awarded to both of them.

Joe asked Darlene, "Okay. Which half do you want? Everyone knows that one divided by two is $\frac{1}{2}$ because I read that in Chapter 15 of *Life of Fred: Fractions.* If you don't mind, I'd like the front half."

Darlene couldn't believe what she was hearing. She yelled at Joe, "Tapirs aren't the same as pizzas!"

Your essay question: How are they different?

4. Fractions only make sense for certain things. In math, we use different numbers for different things. It makes sense to talk about a one-quarter pound of rice but not to talk about one-half of a baby. (King Solomon knew about that in the Old Testament.)

In algebra, we will count backwards from the **whole numbers** {0, 1, 2, 3, 4, . . . } and get **the integers** {. . . –3, –2, –1, 0, 1, 2, 3, . . .}. It doesn't make sense to say that there are –6 people in the room, but it does make sense to talk about –6° when it's six degrees below zero (which happens a lot in the winter in North Dakota).

In advanced algebra, we will talk about numbers that are neither positive (like 5 and 829), nor zero, nor negative (like –24⅓). They are called **imaginary numbers**. Those numbers are not useful for counting babies or measuring temperature, but imaginary numbers are used by engineers when they design airplanes and atom smashers.

Your question: Name six categories (such as tons of steel or volts of electricity) for which fractions make sense.

Complete solutions are on page 176

The Bridge
from Chapters 1–15 to Chapter 16

first try

Goal: Get 9 or more right and you cross the bridge.
Do all the problems before you look at my answers.

Please remember the three General Rules:

(1) Reduce fractions in your answers as much as possible.

(2) Fractions like $\frac{0}{4}$ are equal to 0.

(3) Fractions like $\frac{4}{4}$ are equal to 1.

1. If Betty had carried Fred $\frac{3}{16}$ of the way from his office to the hospital, and Alexander had carried him the rest of the way, what fraction of the trip did Alexander carry Fred?

2. $\frac{7}{20} - \frac{2}{20} = ?$

3. Which is smaller: $\frac{1}{3}$ or $\frac{2}{7}$

4. Write 39 in Roman numerals.

5. Joe was trying to write 49 in Roman numerals. He wrote IL, which was wrong. What should he have written?

6. $\frac{2}{3} + \frac{1}{4} = ?$

7. There is only one whole number that is less than one. Name it.

8. Is $\frac{2}{5} < \frac{1}{3}$ true or false?

9. If you had \$198, how much more would you need to have a million dollars?

10. If there are 360 degrees in a whole circle, how many degrees are in one-fourth of a circle? (360 degrees is sometimes written as 360°.)

The Bridge
from Chapters 1–15 to Chapter 16

second try

1. Joe was given the problem: $\frac{4}{9} - \frac{1}{9} = ?$ and he wrote as his final answer $\frac{3}{9}$. The teacher, who was very strict, marked his answer wrong. Why?
2. Divide LXIV by XVI. Express your answer in Roman numerals.
3. Is $7\frac{1}{3} \geq 7\frac{1}{3}$?
4. If a bottle of basil were $\frac{3}{8}$ full, what fraction of the bottle would be empty?
5. Double 7 eight times. I'll start you off: 7, 14, 28. . . .

6. $\frac{1}{3} + \frac{2}{5} = ?$

7. If sixteen buckets of popcorn are divided equally among 32 friends, how much will each friend receive? (Remember the first General Rule.)
8. If the diameter of a bucket of popcorn is 22", what is its radius?

9. 77 inches is how many feet? Express your answer in feet and inches.

10. Compare $\frac{2}{5}$ and $\frac{1}{2}$ (Which is larger?)

The Bridge

from Chapters 1–15 to Chapter 16

third try

1. $\frac{3}{5} + \frac{3}{8} = ?$

2. Find the sum of LXXIV and XVI. Express your answer in Roman numerals.

3. The garbage can was starting to fill up with empty spice bottles that Fred was throwing away. If it was $\frac{7}{16}$ full, what fraction of the garbage can was empty?

4. Which is smaller: $\frac{1}{3}$ or $\frac{3}{10}$?

5. If there are 360° in a whole circle, how many degrees would be in one-half of a circle?

6. Reduce as much as possible $\frac{204}{276}$ (Hint: When you are done, the numerator will be less than 20.)

7. $\frac{11}{12} - \frac{1}{12} = ?$

8. Write in words 55,444,999.

9. Double 9 eight times. I'll start you off: 9, 18, 36. . . .

10. Joe went fishing with Darlene. Joe told her that in order to attract the fish, he should put some food in the water. He tossed overboard the five baloney sandwiches that Darlene had made. The fifteen ducks that were following the boat shared the sandwiches equally. How much did each duck receive? (Remember the first General Rule.)

The Bridge

from Chapters 1–15 to Chapter 16

fourth try

Please remember the three General Rules:

(1) Reduce fractions in your answers as much as possible.

(2) Fractions like $\frac{0}{4}$ are equal to 0.

(3) Fractions like $\frac{4}{4}$ are equal to 1.

1. Which numeral is larger: 7 or 9 ?

2. $\frac{67}{90} = \frac{?}{720}$

3. On a test, Joe gave an answer of $\frac{6}{8}$ which was marked wrong. Why?

4. $\frac{1}{10} + \frac{1}{40} = ?$

5. Joe wasn't very good at rowing his boat. Once he rowed it in a circle that had a radius of 88 feet. What was the diameter of that circle?

6. Compare $\frac{5}{6}$ and $\frac{6}{7}$

7. Joe told Darlene that his boat was 88 inches long. He thought that made the boat seem longer than if he had expressed it in feet and inches. How long was his boat in feet and inches?

8. A bottle of ketchup is $\frac{5}{8}$ full. What fraction of the bottle is empty?

9. Divide CXXVI by XIV. Express your answer in Roman numerals.

10. If there are 360° in a whole circle, how many degrees would be in one-eighth of a circle?

The Bridge

from Chapters 1–15 to Chapter 16

fifth try

1. If you had \$3,489, how much more would you need to have a billion dollars?

2. If Joe and Darlene were on a date, and Joe ate $\frac{7}{10}$ of the food, how much did he leave for Darlene?

3. Joe likes ketchup. He asked for the extra-large bottle and squirted three cups of it equally on his four potato pancakes. How much ketchup did each pancake receive?

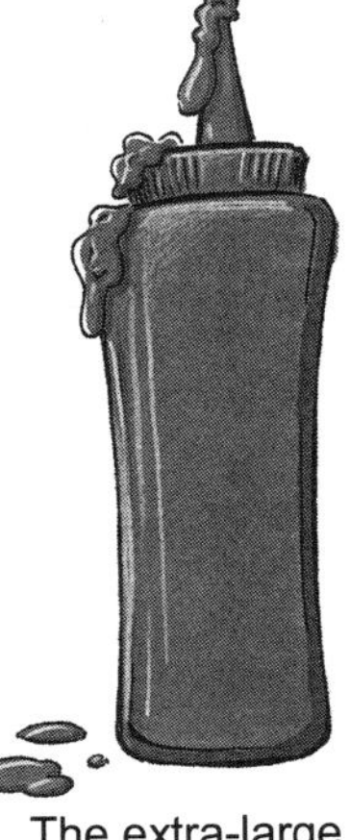

The extra-large size

4. $\frac{87}{90} = \frac{?}{630}$

5. Compare $\frac{2}{3}$ and $\frac{7}{12}$

6. Find the sum of IX and XCIX. Express your answer in Roman numerals.

7. $\frac{1}{9} + \frac{1}{99} = ?$

8. Darlene hoped that some day Joe would propose to her. She was saving money for the wedding. So far, she had saved \$589. How much more would she need to save for a twenty-thousand-dollar wedding?

9. There are two whole numbers that are less than two. Name them.

10. $\frac{11}{12} - \frac{1}{12} = ?$

Chapter Sixteen
Least Common Multiple

Fred asked the eight students if they had any more questions. When they said no, he tied on the dish towel, put on his chef's hat, and headed back to the kitchen.

He was almost done getting the spice bottles straightened out. He was eager to start making pizzas. There were three bottles of onion powder. One had $\frac{3}{4}$ of a tablespoon of onion powder. Another had $\frac{5}{8}$ of a tablespoon, and the third bottle had $\frac{1}{2}$ tablespoon. He poured them all into one of the bottles and threw away the other two.

$$\frac{3}{4} + \frac{5}{8} + \frac{1}{2} = ?$$

We change these fractions so they have a common denominator.

We *could use* 64 as the common denominator. ($64 = 4 \times 8 \times 2$)

Multiply the top and bottom of $\frac{3}{4}$ by 16.

Multiply the top and bottom of $\frac{5}{8}$ by 8.

Multiply the top and bottom of $\frac{1}{2}$ by 32.

We would get:

$$\frac{48}{64} + \frac{40}{64} + \frac{32}{64}$$

but that's way too much work.

We *could use* 16 as the common denominator

since 4, 8, and 2 divide evenly into 16.

Multiply the top and bottom of $\frac{3}{4}$ by 4.

Multiply the top and bottom of $\frac{5}{8}$ by 2.

Multiply the top and bottom of $\frac{1}{2}$ by 8.

We would get:

$$\frac{12}{16} + \frac{10}{16} + \frac{8}{16}$$

but that's way too much work.

Why work so hard? We'll use 8 as the common denominator.

$$\frac{3}{4} + \frac{5}{8} + \frac{1}{2}$$

will become $$\frac{6}{8} + \frac{5}{8} + \frac{4}{8}$$

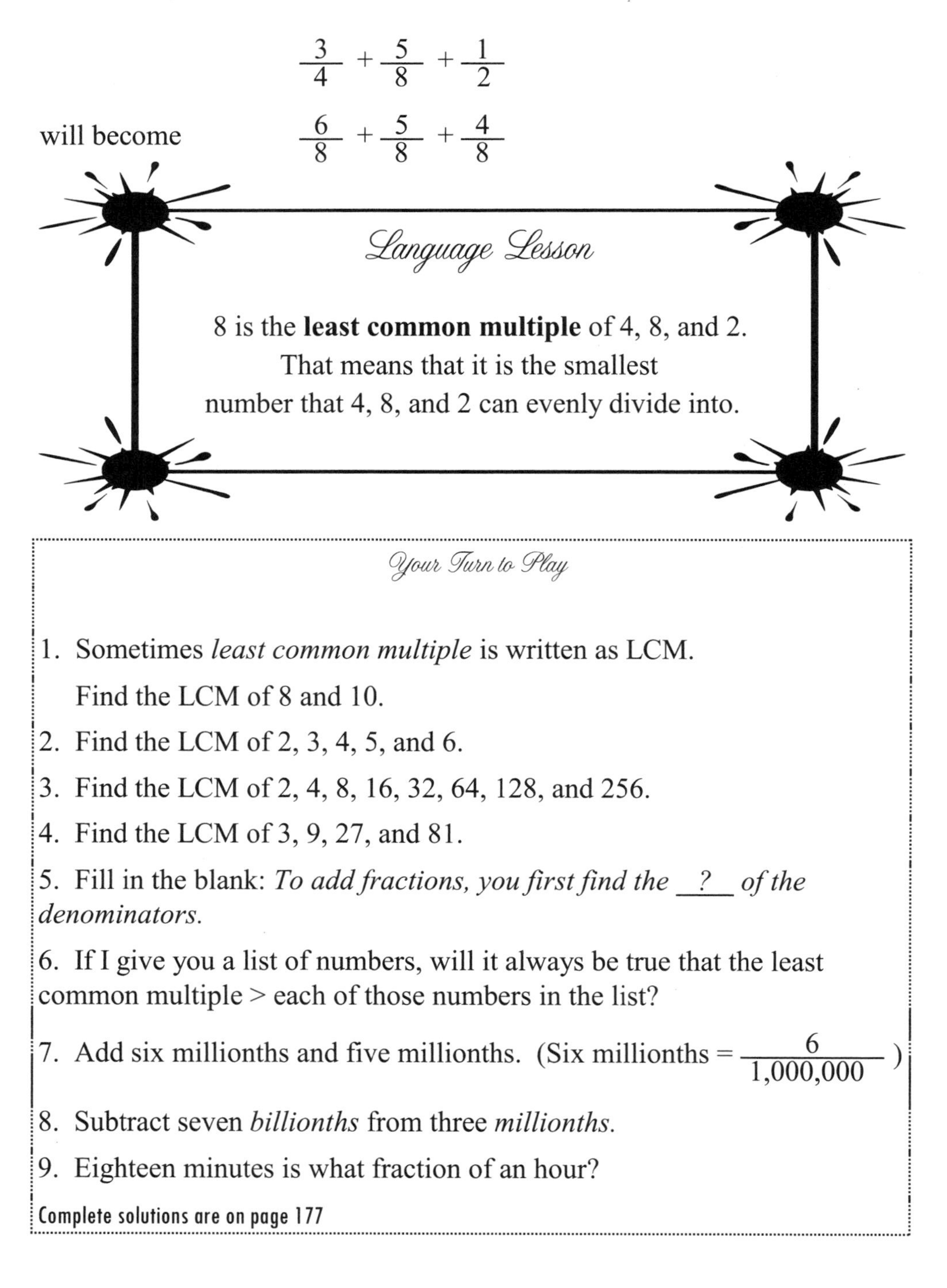

Language Lesson

8 is the **least common multiple** of 4, 8, and 2. That means that it is the smallest number that 4, 8, and 2 can evenly divide into.

Your Turn to Play

1. Sometimes *least common multiple* is written as LCM. Find the LCM of 8 and 10.
2. Find the LCM of 2, 3, 4, 5, and 6.
3. Find the LCM of 2, 4, 8, 16, 32, 64, 128, and 256.
4. Find the LCM of 3, 9, 27, and 81.
5. Fill in the blank: *To add fractions, you first find the _?_ of the denominators.*
6. If I give you a list of numbers, will it always be true that the least common multiple > each of those numbers in the list?
7. Add six millionths and five millionths. (Six millionths = $\frac{6}{1{,}000{,}000}$)
8. Subtract seven *billionths* from three *millionths.*
9. Eighteen minutes is what fraction of an hour?

Complete solutions are on page 177

Chapter Seventeen
Improper Fractions

Fred was in the middle of combining three bottles of onion powder. One had $\frac{3}{4}$ of a tablespoon of onion powder. Another had $\frac{5}{8}$ of a tablespoon, and the third bottle had $\frac{1}{2}$ tablespoon.

He wanted to add $\frac{3}{4} + \frac{5}{8} + \frac{1}{2}$

The least common multiple of 4, 8, and 2 is 8, so the **least common denominator** is 8.

$$\frac{6}{8} + \frac{5}{8} + \frac{4}{8}$$

which equals $\frac{15}{8}$

The numerator is too big. When the top ≥ the bottom, this is called an **improper fraction.***

We know what to do when the numerator = denominator. By one of the General Rules, $\frac{9}{9} = 1$.

What about $\frac{15}{8}$? From Chapter 15, we know that "fractions mean divide."

$$\frac{15}{8} = 8\overline{)15}\ \text{(1 R 7)};\quad 15 - 8 = 7;\quad = 1\frac{7}{8}$$

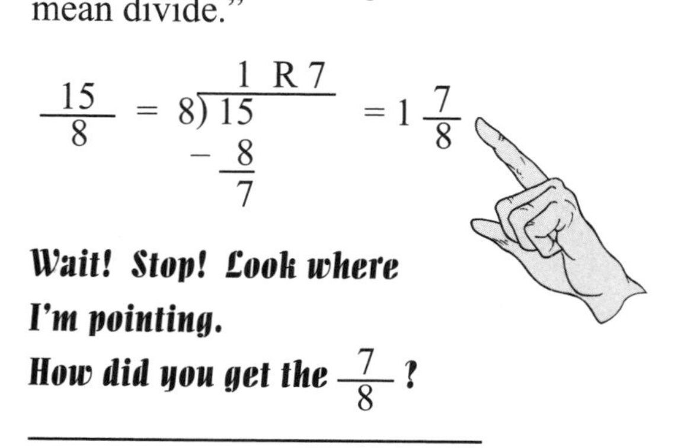

Wait! Stop! Look where I'm pointing.
How did you get the $\frac{7}{8}$?

---

* Who ever uses the word *improper* nowadays? Our great-grandparents might have said that someone "had engaged in improper conduct," but the word is used less frequently today—although a lot of improper conduct is still going on. (Did I just prove that I'm as old as my great-grandparents?)

Good question. I was about to explain that.

When you do a division problem and it doesn't come out evenly, there are two ways to handle it.

<u>Way #1</u>: Express it as a remainder.

Example: Suppose you are asked to convert 54 inches into feet and inches.

$$\begin{array}{r} 4\text{ R }6 \\ 12\overline{)\,54} \\ -\underline{48} \\ 6 \end{array}$$

So 54" = 4' 6"

<u>Way #2</u>: Express the "leftover part" as a fraction.

Example: Suppose you are asked to convert 54 inches into feet.

$$\begin{array}{r} 4\text{ R }6 \\ 12\overline{)\,54} \\ -\underline{48} \\ 6 \end{array}$$

So 54" = $4\frac{6}{12} = 4\frac{1}{2}$ feet

Here's another way of seeing why $\frac{15}{8}$ equals $1\frac{7}{8}$

We could write $\frac{15}{8}$ as $\frac{8}{8} + \frac{7}{8}$

and this is equal to $1 + \frac{7}{8}$ which is the same as $1\frac{7}{8}$

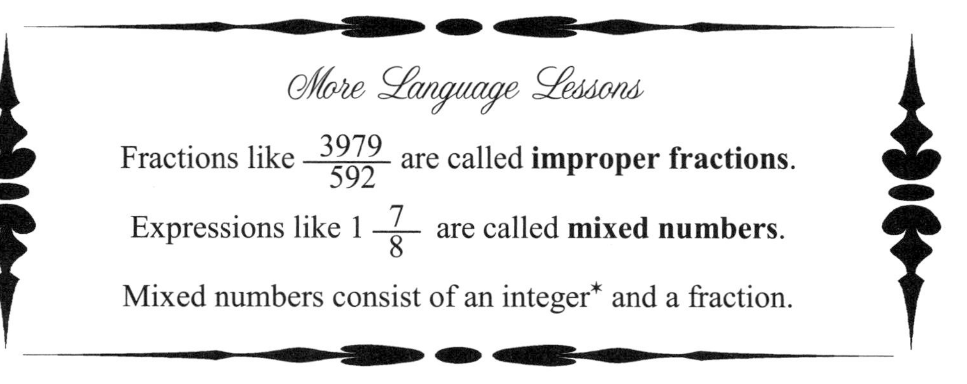

More Language Lessons

Fractions like $\frac{3979}{592}$ are called **improper fractions**.

Expressions like $1\frac{7}{8}$ are called **mixed numbers**.

Mixed numbers consist of an integer* and a fraction.

✶ The integers are the whole numbers {0, 1, 2, 3, 4 . . .} and their negatives. The integers = {. . . –4, –3, –2, –1, 0, 1, 2 . . .}.

Your Turn to Play

1. You know that LCM stands for least common multiple.
Guess what LCD stands for.

2. Fill in the two blanks: *In order to add fractions, you need to find the ____?____ of the fractions—which is the ____?____ of the denominators.*

3. Express in feet and inches: 159".

4. Express in feet: 159".

5. Express in gallons and quarts: 58 quarts. (4 quarts = 1 gallon)

6. Express in gallons: 58 quarts.

Complete solutions are on page 178

The garbage can was filled with empty spice bottles. Fred looked with pride at the neat rows of spice bottles. He didn't know that there were so many different kinds of spices. To make things neat, he had arranged them alphabetically:

Aleppo Pepper, Allspice, Anise, Annatto, Arrowroot, Avocado Leaf, Basil, Bay Leaf, Beet Powder, Bell Pepper, Black Lemon, Caraway, Cardamoms, Celery, Chervil, Chiles, Chives, Cinnamon, Cloves, Cocoa, Coriander, Cream of Tartar, Cubeb, Cumin, Curry Leaf, Dill, Fennel, Fenugreek, Garlic, Ginger, Gumbo File, Hyssop, Juniper, Lavender, Lemon, Lemon Grass, Mace, Marjoram, Mint, Mustard, Nigella, Nutmeg, Onion, Orange, Oregano, Paprika, Parsley, Peppercorns, Poppy, Rosemary, Saffron, Sage, Sesame, Shallots, Star Anise, Sumac, Tarragon, Thyme, Turmeric, Vanilla, and Wasabi.

Now I have elbow room to work he thought to himself. *Now I can be a real chef and create pizzas for PieOne. Stanthony will be so proud of what I have done. Maybe he'll give me a raise.*

"Hey, my dreaming denizen of dining delights.* Where are the pizzas?" Stanthony asked.

Fred smiled. "Just tell me what pizzas you want me to make, and I'll do it in a jiffy.**"

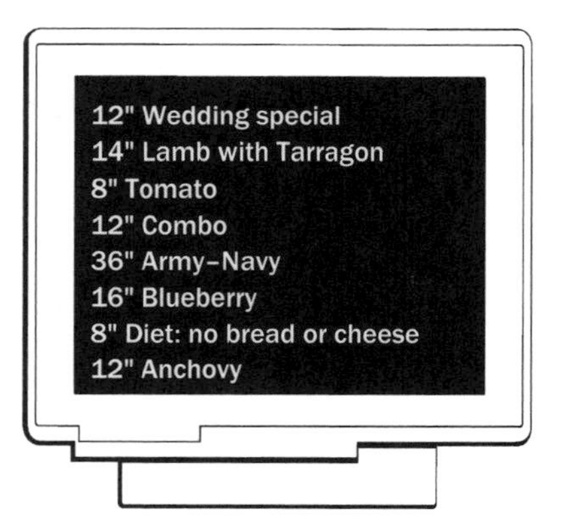

"Tell you?" Stanthony almost shouted. "They are right there on the order screen. What have you been doing?"

Fred blushed. There were a million orders on the screen (hyperbole).

✶ More alliteration by Stanthony. Denizen = a person who lives in a particular place, inhabitant.

✶✶ This is not to be confused with the peanut butter that has a similar name. "In a jiffy" means a very short time. Sometimes this is shortened to "in a jiff."

He looked at the first order:

12" Wedding special

Fred wrinkled his nose. I wonder what that means. Since it's only twelve inches, it's probably just for the bride and groom. I've got to make it very special so that they'll never forget it.

Because for most of his life he had bought his food from the vending machines in the hall down from his office, Fred had never learned much about cooking.

But he did know that the pizza crust was always on the bottom, so he decided he would start there. What shape shall I make the crust? A circle is not very special. I know! I'll make it heart shaped. Fred rolled out some dough and cut it into the shape of a heart.*

pizza crust

A heart will be perfect for a Wedding special pizza. It's a symbol of love. But it's also just the right shape for sharing. No arguing. Cut it right down the middle and the two pieces are exactly alike in shape.

A line of symmetry

With a line of symmetry, one side looks like a mirror image of the other side.

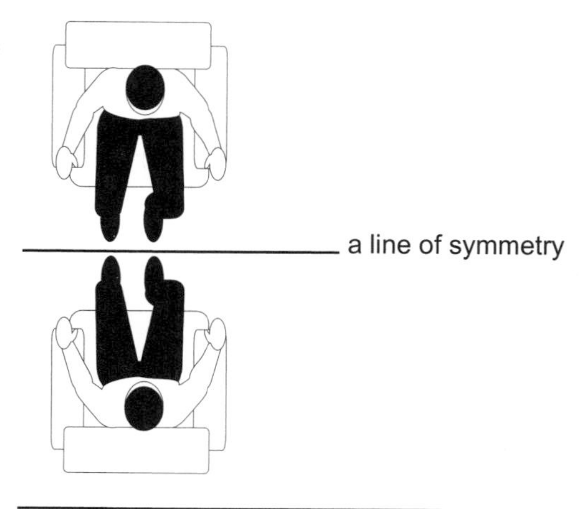

---

* He was *very* careful with the knife.

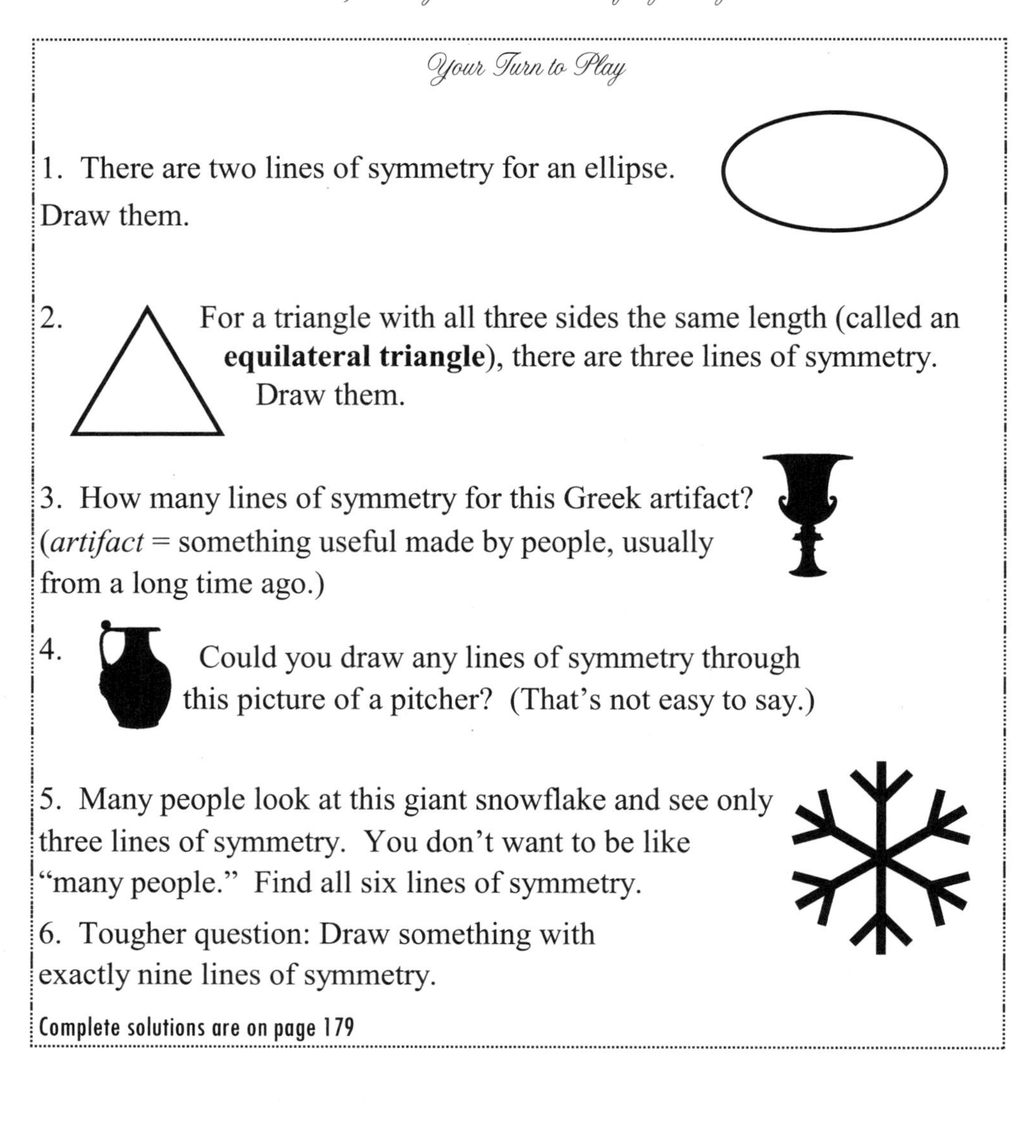

Your Turn to Play

1. There are two lines of symmetry for an ellipse. Draw them.

2. For a triangle with all three sides the same length (called an **equilateral triangle**), there are three lines of symmetry. Draw them.

3. How many lines of symmetry for this Greek artifact? (*artifact* = something useful made by people, usually from a long time ago.)

4. Could you draw any lines of symmetry through this picture of a pitcher? (That's not easy to say.)

5. Many people look at this giant snowflake and see only three lines of symmetry. You don't want to be like "many people." Find all six lines of symmetry.

6. Tougher question: Draw something with exactly nine lines of symmetry.

Complete solutions are on page 179

Chapter Nineteen
Division by Zero

Fred took the heart-shaped pizza crust and smeared it with tomato paste. A red heart. How pretty! I hate to spoil it by putting on toppings that will cover up the red.

Fred skipped the cheese and the toppings and popped his 12" Wedding special pizza into the oven.

Next on the list of pizzas to be made:

14" Lamb with Tarragon

Fred checked in the freezer. There were packages of meat each marked in Stanthony's handwriting: *American bison, steak, veal, yak, hare, rabbit, kangaroo, opossum, goat, ibex, pork, reindeer, deer, moose, antelope, giraffe, squirrel, whale, bear, chicken, duck, dove, quail, mutton, ostrich, pheasant, grouse, partridge, pigeon, woodcock, goose, turkey, frog, anchovy, cod, eel, halibut, salmon, shark, tuna, abalone, clam, conch, mussel, oyster, scallop.* But no lamb.

I wonder if they would accept goose instead? Fred thought to himself. Or maybe I could substitute kangaroo? Do you think they would notice? I'll ask Stanthony. Fred ran out to the front of the restaurant where Stanthony was talking with some of his customers.

"Excuse me," Fred began. "I couldn't find any lamb for the 14" Lamb with Tarragon pizza. Where do you keep it?"

Stanthony pointed back toward the kitchen at the rear of the store while he continued talking with his customers.

Fred didn't know what to do. He just started walking in the direction that Stanthony had pointed.

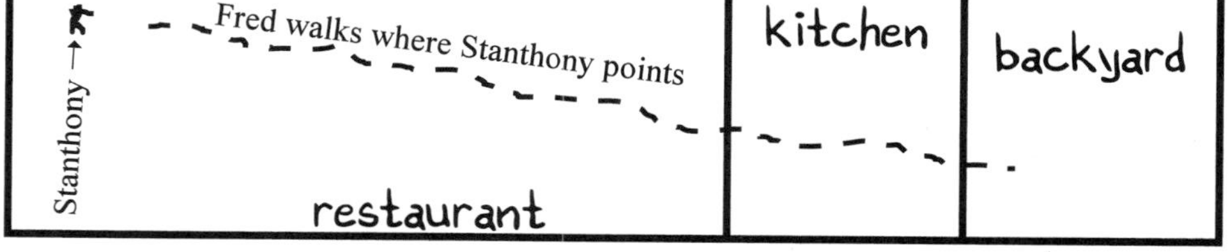

Stanthony →
Fred walks where Stanthony points
restaurant
kitchen
backyard

He kept walking. He walked through the kitchen and out into the backyard. Sometimes in the life of a five-year-old, things don't seem to make sense.* How am I going to make lamb pizza? The kitchen doesn't have any lamb in the freezer. Teaching math is so much easier than making pizza. I need to keep this job since I spent all the money in my checking account on my new bicycle.

Fred's eyes were half-closed as he walked and talked to himself. When he came to the fence in the backyard, he stopped and turned around. He decided that he was going to go back and ask Stanthony again.

He heard a sound. It sounded like "baa." He opened his eyes.

Fred tried to scream—but couldn't. He tried to laugh—but nothing came out. The shock was too great—he passed out.

* This is also sometimes true for people who are 8. Or 11. Or 14. Or 18. Or 26. Or 35. Or 44. Or 48. Or 59. Or 60. Or 74. Or 81. Things sometimes just don't make sense. When God told Abram (also known as Abraham) that he was going to be a daddy when he was 100 years old, he had a little trouble making sense of that. He did what many people would do in that situation: he fell down and laughed (Genesis 17:17). And four chapters later, along comes his newborn son whom they named Giggles. In the original language, the son was called Isaac, which can be translated as *he laughs.*

Your Turn to Play

1. Fred knew he was being asked to do something that he really couldn't do. He couldn't convert into lamb chops.

In mathematics, there are things that are just as hard to do.

Your question: Find a value for the question mark in: $0 \times ? = 14$.

2. How many 3's do you have to add up to get 21?

3. How do you write *How many 3's do you have to add up to get 21?* as a fraction?

4. Using the above question 3 as an example, express $\frac{20}{5}$ as a question.

5. Express $\frac{6}{1}$ as a question.

6. Express $\frac{14}{0}$ as a question.

Complete solutions are on page 179

The Bridge

from Chapters 1–19 to Chapter 20

first try

Goal: Get 9 or more right and you cross the bridge.

1. Express a million feet in miles and feet. (1 mile = 5,280 feet)
2. What is the LCM of 6 and 8?
3. This little lamb is the *first* real live lamb that Fred had ever seen at Stanthony's. Is *first* a cardinal number?

4. Which of these five symbols

   $<, \leq, =, >,$ or $\geq$

   is the best one to fill in the blank:

   An improper fraction is one in which the denominator is __?__ the numerator.

5. If Alexander were seven times richer, he would have $5,523. How much money does he have?

6. This is a nine-pointed blob. How many lines of symmetry does it have?

7. How many lines of symmetry does a circle have?
8. How many 8's do you have to add up to get 6,232?
9. Write out in words: 7,000,944,333.
10. Suppose we have a number which we will call y. Is it possible that both $y \leq 44$ and $y < 62$ are true?

The Bridge

from Chapters 1–19 to Chapter 20

second try

1. 62 quarts are how many gallons? (4 quarts = 1 gallon)
2. Give an example of a mixed number.
3. $\frac{4}{6} + \frac{4}{6} + \frac{4}{6} = ?$ Do not leave your answer as an improper fraction.
4. Draw something with exactly seven lines of symmetry.
5. Express 1000 fluid ounces in cups. (8 fluid ounces = 1 cup)
6. How many lines of symmetry does a sector have?
7. Each month, Joe spends $17 on jelly beans. How much does he spend in a year?
8. Write in Roman numerals 59 and 102.
9. What is the LCM of 8 and 10?
10. How many lines of symmetry does this parallelogram have?

A parallelogram is a four-sided figure in which the opposite sides are parallel.

The Bridge

from Chapters 1–19 to Chapter 20

third try

1. $\frac{7}{8} + \frac{7}{9} = ?$ Do not leave your answer as an improper fraction.

2. If Fred's bicycle box were 8 times heavier, it would weigh 632 pounds. How much does it weigh?

3. Divide DCXXIII by VII. Express your answer in Roman numerals.

4. What is the LCM of a million and a billion?

5. Reduce as much as possible $\frac{48}{54}$

6. If the numerator of a fraction is greater than or equal to the denominator, what do we call such a fraction?

7. Express 7 miles in feet. (1 mile = 5,280 feet)

8. Write out in words: 8,247,921.

9. Name the smallest ordinal number.

10. How many lines of symmetry does this trapezoid have?

(A **trapezoid** is a four-sided figure with exactly two sides parallel.)

The Bridge

from Chapters 1–19 to Chapter 20

fourth try

1. How many lines of symmetry does this diagram have?

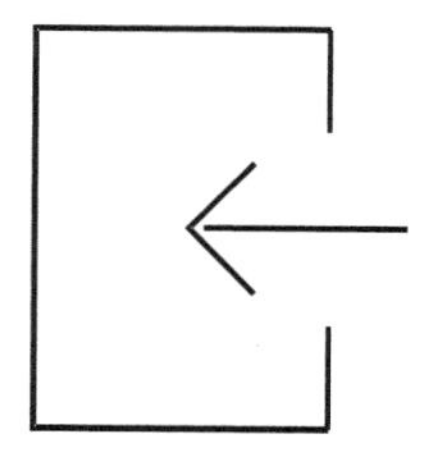

2. What is the LCM of 20 and 20,000?
3. Suppose we have a number which we will call z. Is it possible that both $z \geq 100$ and $z < 50$ are true?
4. Suppose x is some integer.

   (The integers are { . . . –3, –2, –1, 0, 1, 2, 3, 4, 5, 6 . . .})

   There are three possible values of x that make $7 \leq x \leq 9$ true.

   Name them.
5. How many 7's would you have to add up to get 62,629?
6. What is the LCD of $\frac{1}{6}$ and $\frac{1}{8}$?
7. Express in feet and inches: 111"
8. Express in feet: 111"
9. Reduce as much as possible $\frac{50}{90}$
10. Express a million minutes in hours and minutes.

The Bridge

from Chapters 1–19 to Chapter 20

fifth try

---

1. Six bags of jelly beans weigh 534 ounces. How much does one bag weigh? (These are the extra large bags that Joe likes to buy. He can eat a whole bag during one of Fred's lectures. The bags all weigh the same.)

2. What is the radius of a circle that has a 50-furlong diameter?

3. $\frac{7}{8} + \frac{6}{8} + \frac{5}{8} = ?$ (Do not leave your answer as an improper fraction.)

4. Multiply VII by LXXVII. Express your answer in Roman numerals.

5. Express 3,157 quarts in gallons and quarts. (4 quarts = 1 gallon)

6. How many lines of symmetry does this have?

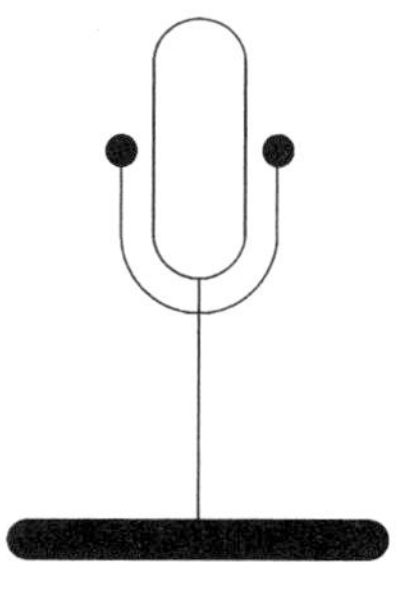

7. Write as a number: Thirty-three billion, nine hundred twenty-six.

8. Does $\frac{0}{2}$ make sense? If it does, what does it equal? If it doesn't, explain why it doesn't.

9. What is the LCM of 3, 4, 5, and 6?

10. What is the LCD of $\frac{1}{4}$, $\frac{1}{5}$, and $\frac{1}{8}$?

Chapter Twenty
Subtracting Fractions

The little lamb thought that Fred was playing with her. She came over and jumped on Fred's chest. When she jumped up and down, she could feel little puffs of air coming out of Fred's nose. She thought that was funny.

Fred awoke. The lamb said, "Hi!"*

Fred said, "Baa!"**

The lamb jumped over Fred's nose—a very dangerous thing to do, given how sharp his nose is. She went and ate the two flowers.

Stanthony came out into the backyard and saw Fred lying on the ground.*** "What kind of a worker are you?" Stanthony roared. "The customers are waiting for their 14" Lamb with Tarragon, and you're lying out here in the backyard looking at the sky."

Little lamb came over to Stanthony and rubbed up against his leg. "And you were playing with my pet lamb," he continued.

"Your pet lamb?" Fred asked. "And you want me to kill her to make somebody's pizza?"

"Where did you get that idea?"

"Well," said Fred. "When I asked you where you keep the lamb for the pizza, you pointed to the backyard."

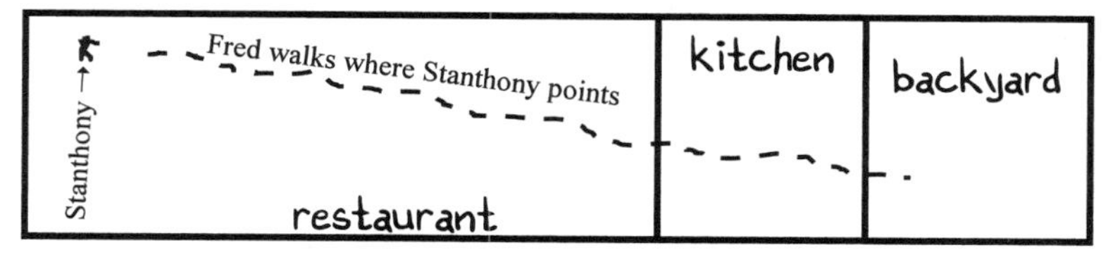

✶ which in lamb-talk, it sounds like "Baa!"

✶✶ which in people-talk means "Hi!"

✶✶✶ He was *lying* on the ground—not *laying*. To *lay* is to set something down. Chickens lay eggs on the ground. If someone tells you to, "Go lay down," you can ask them, "What do you want me to lay down?" They should have said, "Go lie down." It's gooder English.

Stanthony laughed. "No, you silly. I pointed to the kitchen. We use the mutton for our lamb pizza." He patted little lamb on the head.

Fred thought back over packages of meat: *American bison, steak, veal, yak, hare, rabbit, kangaroo, opossum, goat, ibex, pork, reindeer, deer, moose, antelope, giraffe, squirrel, whale, bear, chicken, duck, dove, quail, mutton, ostrich, pheasant, grouse, partridge, pigeon, woodcock, goose, turkey, frog, anchovy, cod, eel, halibut, salmon, shark, tuna, abalone, clam, conch, mussel, oyster, scallop.*

There it was. How could I have missed that? Mutton is the meat from grownup lambs—from mature sheep. I am so glad. I am so glad. I am so glad. He hopped to his feet and ran into the kitchen.

Fred was so happy. He made up a little song and sang it:

Stannie had a little lamb,
Its fleece was white as snow,
And all the time I thought she'd be
A topping on the pizza dough.

He pulled the package marked mutton out of the freezer.

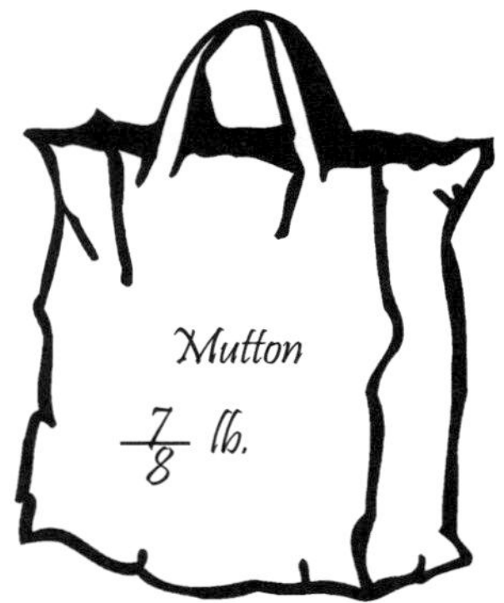

The abbreviation for pound is lb.

Wait! Stop! I, your reader, don't understand that. Why should lb. be the abbreviation for pound? Shouldn't it be something like "po." or "pd."?

The ancient Romans had a unit of weight called a *libra* (pronounced LIE-breh). *Libra* is the Latin word for "scales."* When we translated it into English, we kept the Latin abbreviation.

Is that why we abbreviate ounces as oz.?

You are almost right. In Latin, "ounces" is *uncia.* That became the Italian word *onza.* In the 1400's, England did a lot of trading of wool with Italy, and it's from the Italian that we get "oz."

And the British symbol £ for their currency is called the "pound"?

Yes. £12 means 12 pounds and it's from the Latin *libra.*

✶ The constellation in the sky called Libra is pronounced LEE-breh.

Fred chopped off a quarter of a pound of mutton. He crossed off the $\frac{7}{8}$ lb. and wrote $\frac{5}{8}$ lb. on the package.

Your Turn to Play

1. Fred took $\frac{7}{8}$ lb. and chopped off $\frac{1}{4}$ lb. and had $\frac{5}{8}$ lb. left. That's subtraction. Show how $\frac{7}{8} - \frac{1}{4} = \frac{5}{8}$

2. $\frac{1}{3} - \frac{1}{4} = ?$

3. If Fred had started with $\frac{7}{8}$ lb. of mutton and had chopped off $\frac{1}{5}$ lb., how much would have been left?

4. In one-half hour of employment, he spent one-third hour sorting spice bottles in the kitchen and the rest of the time with little lamb in the backyard. What fraction of an hour did Fred spend in the backyard?

5. One-sixth of an hour is how many minutes?

6. Fred was being paid \$12/hour by Stanthony. How much did he "earn" during the $\frac{1}{6}$ hour he was in the backyard?

7. Each time little lamb jumped on Fred's chest, she made four hoof marks on him. If she jumped 17 times, how many hoof marks did Fred get?

8. Trick question: If it takes 3 days for a hoof mark to disappear from Fred's chest, how long will it take for the 68 hoof marks to disappear?

9. Stanthony bought four identical shoes for his pet lamb. The four shoes weighed a total of 3 lbs. How many pounds did each shoe weigh?

10. When Joe went out fishing with Darlene, she packed two baskets. One of them was their lunch—five balogna sandwiches. The other basket had the bait in it—zillions of little meal worms. Joe accidentally ate $\frac{6}{17}$ of the basket with the bait. What fraction of the bait basket did he leave uneaten?

Complete solutions are on page 180

Chapter Twenty-one
Circumference

Fred rolled out the pizza dough for the 14" Lamb with Tarragon. When the diameter of the circle of crust was exactly 14", he smeared on the red pizza sauce. He took the quarter pound of mutton and ran it through the grinder so that it would look more like hamburger than like steak.

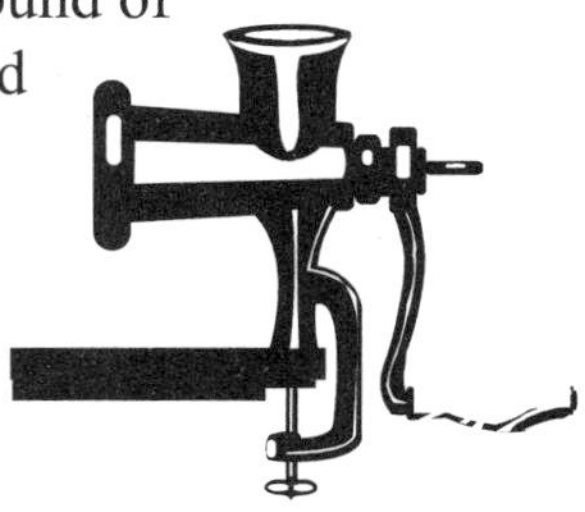

Now it's time to be artistic Fred thought to himself. He took little pieces of "lamb burger" and put them around the edge of the crust.

The pieces were on the **circumference** of the circle. Fred then took the pieces off the pizza and put them all in a straight line and measured the length.

It was about 44". It was a little more than three times the diameter.

Actually, it was closer to $3\frac{1}{7}$, not exactly equal to $3\frac{1}{7}$ but pretty close.

14" times $3\frac{1}{7}$ gives you the circumference of a 14" pizza.

Wait! I know how to add fractions and subtract fractions, but we have never multiplied fractions. Did you forget to explain how to do that?

No I didn't. I was just about to do that when you interrupted.

Okay. Explain how to multiply fractions.

It's much easier than adding fractions. Just look at this example, and you will know how to do it.

$$\frac{5}{8} \times \frac{3}{7} = \frac{15}{56}$$

Hey! That's too easy. You just multiply top times top and bottom times bottom.

You took the words right out of my mouth. Multiplying fractions is a lot easier than adding fractions. You don't have to find any least common denominator.

Here is the world's shortest *Your Turn to Play*:

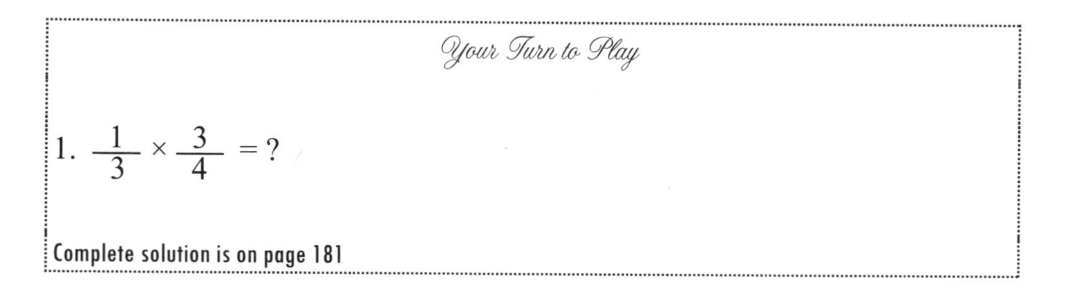

Your Turn to Play

1. $\frac{1}{3} \times \frac{3}{4} = ?$

Complete solution is on page 181

Fred placed ***Hey! Wait! Stop! Don't start this chapter yet. I've got some questions left over from Chapter 21. You did a nice job of explaining how to multiply fractions: top times top and bottom times bottom. I'm grateful that you didn't give me 4000 problems to do. I hate to spend all afternoon doing . . .***

Homework:

1. 1/3 × 1/4
2. 2/5 × 1/7
3. 3/8 × 4/9
4. 2/9 × 3/4
5. 2/7 × 2/9
6. 3/5 × 9/11
7. 5/6 × 1/7
8. 8/9 × 1/3
9. 4/7 × 3/4
10. 3/8 × 7/9
11. 2/5 × 3/7
12. 1/11 × 2/3
13. 4/5 × 3/10
14. 3/5 × 7/9
15. 1/3 × 2/5
16. 2/9 × 2/5
17. 7/8 × 3/4
18. 3/8 × 2/7
19. 4/5 × 1/8
20. 5/6 × 6/7
21. 6/7 × 7/8
22. 7/8 × 8/9
23. 3/8 × 1/2
24. 2/7 × 3/5
25. 3/7 × 4/9
26. 7/9 × 6/3
27. 3/8 × 2/9
28. 3/7 × 7/8
29. 1/8 × 1/9
30. 2/5 × 5/6

But you talked about multiplying 14" times $3\frac{1}{7}$***, but you never showed me how to do that.***

Okay. $3\frac{1}{7}$ is a mixed number. You're right. I showed you how to multiply fractions, but I didn't talk about multiplying mixed numbers.

Back in Chapter 17, we changed improper fractions into mixed numbers.

We started with $\frac{15}{8}$, did this: $8\overline{)15}$ with $1\text{ R }7$ on top, and got $1\frac{7}{8}$

Now we'll reverse the process. We will change mixed numbers into improper fractions.

Here is the **hard way** to change mixed numbers into improper fractions:

$$3\frac{1}{7} = \frac{3 \times 7}{1 \times 7} + \frac{1}{7} = \frac{21}{7} + \frac{1}{7} = \frac{22}{7}$$

Your Turn to Play

1. Change $5\frac{1}{6}$ to an improper fraction.
2. Change $11\frac{1}{10}$ to an improper fraction.
3. Change $2\frac{5}{8}$ to an improper fraction.

Complete solutions are on page 181

And then there is the **easy way** to change mixed numbers into improper fractions.

Suppose you start with $3\frac{1}{7}$ and you get to $\frac{22}{7}$

The words you say in your head are: *7 times 3 . . . plus 1.*

If you start with $8\frac{2}{5}$ you say: *5 times 8 . . . plus 2,* and you get $\frac{42}{5}$

Now we're going to repeat the *Your Turn to Play* that we just did, but we'll use the easy way.

Your Turn to Play

1. Change $5\frac{1}{6}$ to an improper fraction.
2. Change $11\frac{1}{10}$ to an improper fraction.
3. Change $2\frac{5}{8}$ to an improper fraction.

Complete solutions are on page 181

Wait! Stop! My question at the beginning of this chapter was about multiplying 14" times $3\frac{1}{7}$ and all you have been talking about is changing mixed numbers into improper fractions.

I've been trying to answer your question. If you want to multiply mixed numbers, you first change them into improper fractions.

$14 \times 3\frac{1}{7}$ becomes $\frac{14}{1} \times \frac{22}{7}$ and the rest is "old stuff" from the previous chapter. Namely, you multiply top times top and bottom times bottom.

Your Turn to Play

1. If we have a pizza with a diameter of 14", and we multiply that by $3\frac{1}{7}$ to get the approximate circumference, what will be the answer?

Complete solution is on page 181

Chapter Twenty-three
Commutative Law

So Fred placed about 44" of lamb burger around the circumference of his 14" Lamb with Tarragon pizza. With the rest of his quarter pound of mutton he did what any five-and-a-half-year-old would do.

He sprinkled some tarragon over the happy face. That looks like freckles, he thought to himself. He set the pizza aside and started work on:

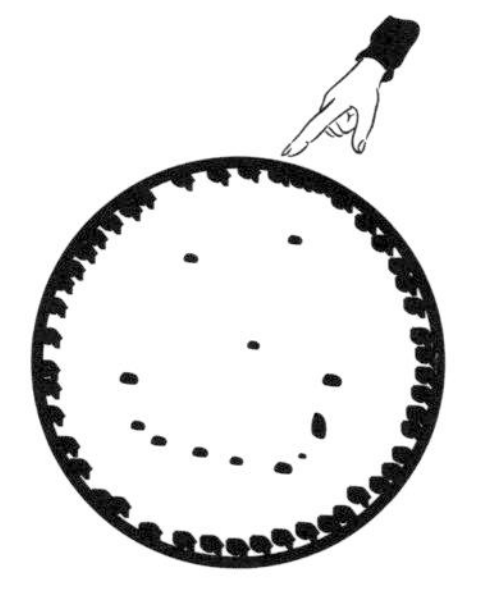

8" Tomato

He took a couple of Roma tomatoes* and sliced them. Do I put them on and then cook the pizza—or do I cook the pizza and then put them on? he wondered. Those operations are not commutative. It makes a difference which order I do them in.

Addition is **commutative**. Adding 3 + 7 gives you the same answer as 7 + 3. (In algebra we say that the commutative law for addition is a + b = b + a for any two numbers, a and b.)

When Fred first learned about the commutative law of addition, he said, "When I stick 3 jelly beans in a bag and then put in 7 more, I get the same as when I stick 7 jelly beans in a bag and then put in 3 more."

Fred looked around the kitchen and played with the idea of which things were commutative and which weren't.

1. If I set the clock and open a can of sauce, I get the same result as when I open a can of sauce and then set the clock. Commutative.

2. If I open a can of pizza sauce and then pour it on the pizza—I can't do that in the other order! It's really not commutative.

3. If I put a tissue to my nose and then blow it, I get a really different result if I reverse the order. Not commutative.

---

* Stanthony would have said red, ripe, Roma. Alliteration.

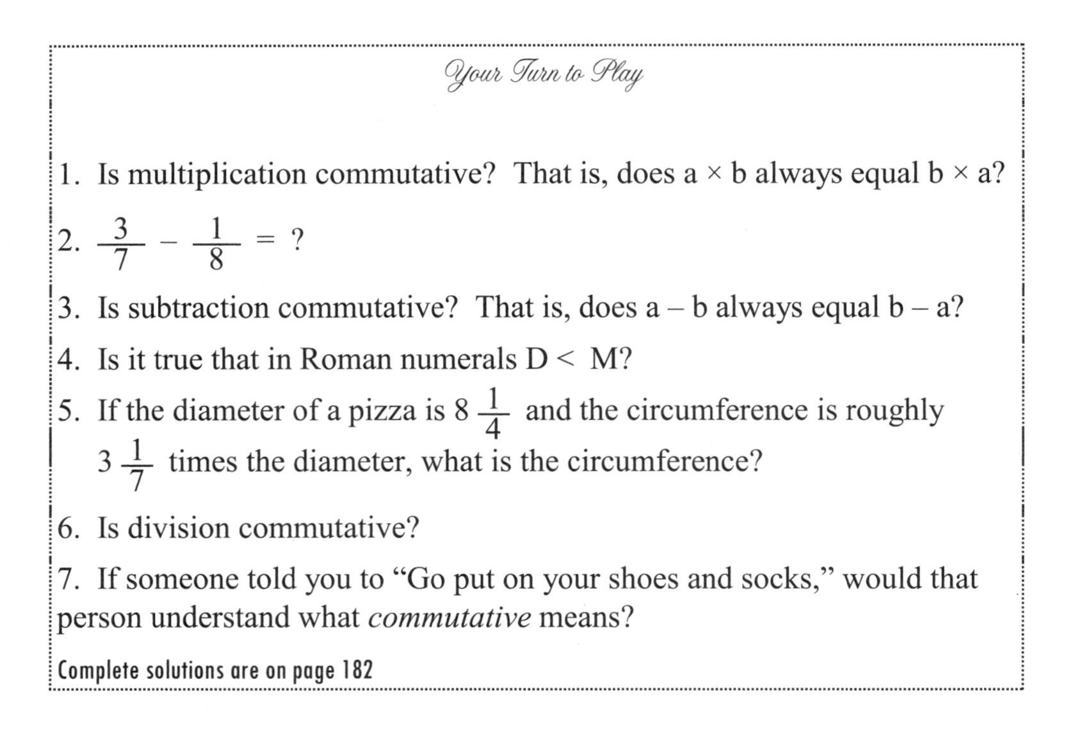

Your Turn to Play

1. Is multiplication commutative? That is, does a × b always equal b × a?
2. $\frac{3}{7} - \frac{1}{8} = ?$
3. Is subtraction commutative? That is, does a – b always equal b – a?
4. Is it true that in Roman numerals $D < M$?
5. If the diameter of a pizza is $8\frac{1}{4}$ and the circumference is roughly $3\frac{1}{7}$ times the diameter, what is the circumference?
6. Is division commutative?
7. If someone told you to "Go put on your shoes and socks," would that person understand what *commutative* means?

Complete solutions are on page 182

Chapter Twenty-four
Adding Mixed Numbers

Fred hadn't read the instruction book: Prof. Eldwood's *Guide to Making Pizza*, 1851. It gives all the steps:

1. Make crust.
2. Smear with pizza sauce.
3. Add toppings and cheese.
4. Stick in oven.
5. Take out of oven.

Fred had missed the last step. Five chapters ago, he had "popped his 12" Wedding special pizza into the oven." That was a long time ago.

First, it cooked.

Then it dried out.

Then it blackened.

Then it caught on fire.*

Smoke poured out of the oven. The kitchen filled with smoke. Everyone in the restaurant started coughing. Even little lamb in the backyard was affected.

The smoke alarm went off.

The sprinkler system showered everyone.

The fire engines rolled up in front of PieOne.

And Fred was out of a job.

The reporters from the KITTEN Caboodle newspaper arrived. They interviewed Stanthony. They talked with the wet customers. And everyone was pointing at Fred.

The federal disaster-relief group arrived. Six members of Congress were with them. All six got their picture taken with little lamb. One

---

* Did Smokey the Bear ever say, "Things burn better when they're dried out"?

disaster worker handed Stanthony a package containing a moist little paper towel and told him it was for the lamb. Then the disaster relief group and the members of Congress left.

Fred didn't know what to do. He walked slowly down the street in a daze. The special edition of the newspaper was already published.

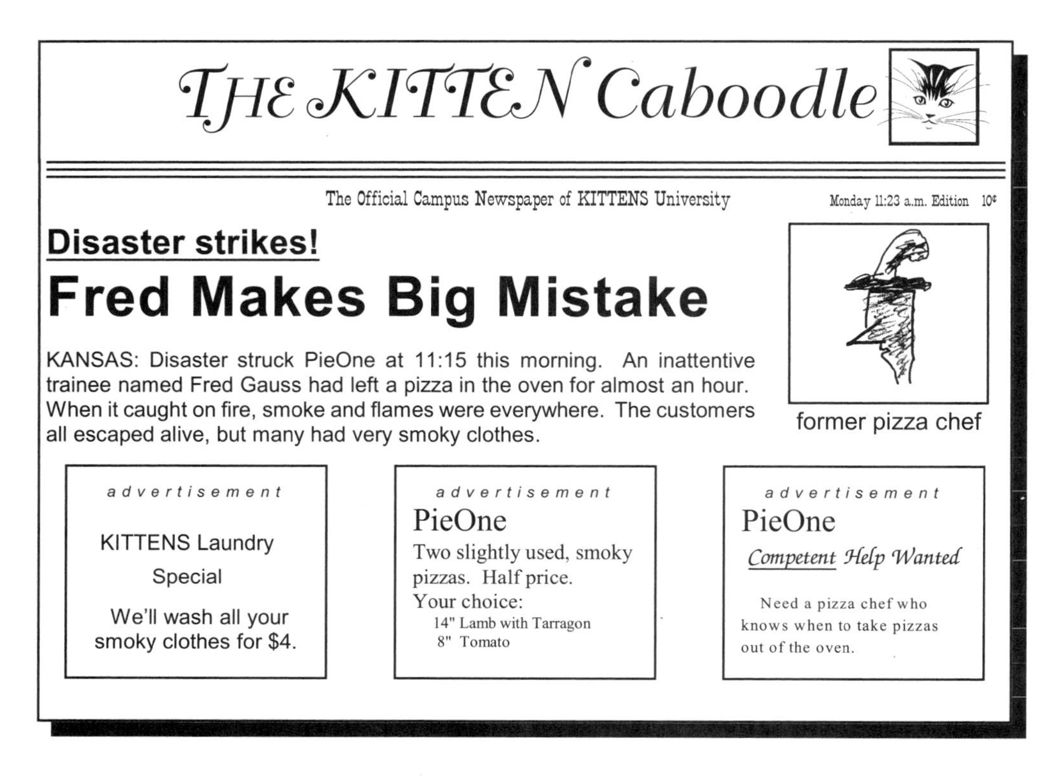

THE KITTEN Caboodle

The Official Campus Newspaper of KITTENS University — Monday 11:23 a.m. Edition 10¢

Disaster strikes!

Fred Makes Big Mistake

KANSAS: Disaster struck PieOne at 11:15 this morning. An inattentive trainee named Fred Gauss had left a pizza in the oven for almost an hour. When it caught on fire, smoke and flames were everywhere. The customers all escaped alive, but many had very smoky clothes.

former pizza chef

advertisement

KITTENS Laundry
Special
We'll wash all your smoky clothes for $4.

advertisement

PieOne
Two slightly used, smoky pizzas. Half price.
Your choice:
14" Lamb with Tarragon
8" Tomato

advertisement

PieOne
Competent Help Wanted
Need a pizza chef who knows when to take pizzas out of the oven.

What to do? I don't have any money in my checking account. I'm out of a job, except, of course, for my teaching at KITTENS. But my paycheck for teaching comes at the end of the month.

Fred put his hands in his pockets as he walked down the street. This wasn't the usual way that he walked. He didn't even feel like singing.

Like many five-and-a-half-year-old boys, his pockets were filled with stuff. He could hardly get his hands in his pockets. He stopped at a bench and emptied his pockets onto the bench. Out of his left pocket he pulled $3\frac{3}{4}$ sticks of gum. Out of his right pocket, $2\frac{1}{2}$ sticks of gum.

Well, at least I won't starve he thought to himself.

Fred had never been what you would call a "big eater." Normally, he would have a quarter of a stick of gum. A half stick of gum was a really big mouthful for him.

Your Turn to Play

1. How much gum did Fred have?

We need to add $3\frac{3}{4}$ and $2\frac{1}{2}$

That's done by adding the whole numbers (the 3 and the 2) and then adding the fractions ($\frac{3}{4}$ and $\frac{1}{2}$).

That's how you add mixed numbers.

2. Then Fred pulled out a bag of nuts that weighed $1\frac{1}{2}$ oz. He found another bag of nuts that weighed $2\frac{1}{3}$ oz., and another bag of nuts and bolts that weighed $4\frac{2}{3}$ oz. Fred was hoping to someday build a robot, so he had been collecting anything that might be of help. He combined his three bags into one bag. How much did it weigh?

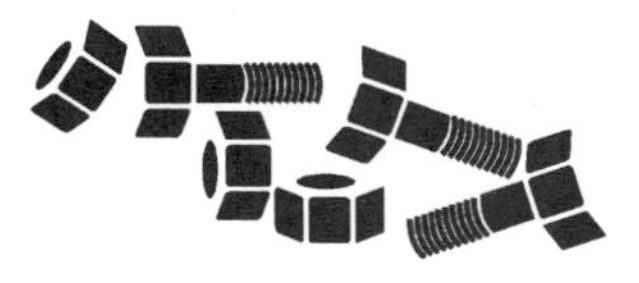

3. $89\frac{1}{7} + 2\frac{1}{7} = ?$

4. $89\frac{1}{7} \times 2\frac{1}{7} = ?$

5. $89\frac{1}{7} - 2\frac{1}{7} = ?$

6. $0 \times 3{,}976{,}988 = ?$

7. $0 \times 0 = ?$

8. $89\frac{1}{7} + 0 = ?$

9. Which is smaller: $\frac{2}{5}$ or $\frac{3}{7}$?

10. Down underneath the bags of nuts and bolts, Fred found some coins: 42 pennies, 27 nickels, 3 dimes, and 6 quarters.

How much money did he have? Express your answer in cents.

Complete solutions are on page 182

The Bridge

from Chapters 1–24 to Chapter 25

first try

Goal: Get 9 or more right and you cross the bridge.

1. If little lamb could jump up and down 17 times each minute, how many times could she jump up and down in an hour?

2. If little lamb eats $15\frac{1}{7}$ ounces of food each day, how many ounces of food would she eat in a week?

3. If I asked you to name an integer (= { . . . –3, –2, –1, 0, 1, 2, 3 . . .}) that would make $5 < x < 6$ true, you couldn't do it.

   Instead, I ask you to find some number y which makes $5 < y < 6$ true.

4. What is the LCM of 7, 14, and 28?

5. Change $\frac{1515}{17}$ to a mixed number.

6. Is it possible to name an integer that is not a whole number? If so, give an example. If not, explain why it is not possible.

7. Joe weighed $148\frac{1}{2}$ lbs. before he went out to have lunch with Darlene. He ate $1\frac{7}{8}$ lbs. of food at lunch. How much did he weigh after lunch?

8. A regular monster burger has $\frac{5}{8}$ lb. of meat in it. Joe ordered a double monster burger. How much meat was in it?

9. For this problem, we'll say that the circumference of a circle is equal to $3\frac{1}{7}$ times the diameter. What would be the circumference of a bicycle tire whose ***radius*** is equal to $8\frac{1}{4}$ inches?

10. What number times a million would give you a billion?

The Bridge

from Chapters 1–24 to Chapter 25

second try

1. Convert $73\frac{1}{3}$ to an improper fraction.

2. Stanthony put $8\frac{7}{8}$ oz. of lamb food in little lamb's bowl. She ate $4\frac{2}{3}$ oz. How much was left in her bowl?

3. Reduce as much as possible $\frac{42}{54}$

4. On Joe's plate was a $\frac{3}{4}$ lb. hamburger, $\frac{1}{8}$ lb. of French fries, and $\frac{1}{2}$ lb. of ketchup. How much did all that weigh?

5. Write in Roman numerals the numbers from 21 to 30.

6. Find the LCM of 6 and 9.

7. Draw something that has exactly two lines of symmetry.

8. Write out in words: 57,983,000,000.

9. Suppose you have x dollars in your pocket and you want to buy an oboe that costs $143. Which is better? $x < 143$ or $x > 143$.

Made of wood. Looks a little like a clarinet, but sounds quite different. The oboe uses a double reed, and many oboists make their own reeds. Many times, sad melodies are played by the oboe in an orchestra.

pronounced
OH-beau

I was going to write
OH-bow
but there are two different *bow* words:
1. bow and arrow
and
2. bow down.

10. Change $\frac{992}{34}$ to a mixed number.

The Bridge

from Chapters 1–24 to Chapter 25

third try

1. Once, for lunch Fred had $\frac{1}{4}$ oz. of lettuce, $\frac{1}{8}$ oz. of hamburger, and $2\frac{1}{2}$ oz. of onion. (He was really in the mood for onions.) How many ounces of food did he eat?

2. For this problem, we'll say that the circumference of a circle is equal to $3\frac{1}{7}$ times the diameter. What would be the circumference of a circle whose diameter is equal to $11\frac{1}{10}$ feet?

3. $55\frac{1}{8} + 27\frac{3}{4} = ?$

4. Cardinal or ordinal? When Fred pulled the third bag of nuts and bolts out of his pocket, he knew that *third* is a(n) ____________ number.

5. Cardinal or ordinal? When Fred pulled the third bag of nuts and bolts out of his pocket, there were 4 nuts in it. He knew that 4 is a(n) ____________ number.

6. $\frac{3}{4} \times \frac{7}{6} = ?$

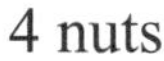

4 nuts

7. If little lamb says "Baa!" five times each minute, how many times would she say "Baa!" in six hours?

8. $\frac{6}{7} - \frac{2}{3} = ?$

9. Divide CXXXVI by XVII and give your answer in Roman numerals.

10. Change $\frac{276}{38}$ to a mixed number.

The Bridge

from Chapters 1–24 to Chapter 25

fourth try

1. Change $\frac{287}{71}$ to a mixed number.

2. When Darlene was getting ready to go fishing with Joe, she needed to do two things: pack some bait in her purse in case Joe forgot to bring some, and, second, put on some sun screen lotion so she wouldn't get sunburned. Are those two things—packing bait and putting on sun screen—commutative?

3. When Joe and Darlene got out in the boat, she asked him if he had remembered to use sun screen. He hadn't. She offered him her bottle. It had $6\frac{2}{3}$ ounces of sun screen in it. Joe used $5\frac{1}{2}$ ounces. How much did he leave in the bottle?

4. Little lamb can run 278" in a minute. Convert that to feet and inches.

5. Convert $1000\frac{2}{3}$ to an improper fraction.

6. What is the LCM of 20 and 30?

7. Draw something that has exactly three lines of symmetry.

8. There are 360° in a circle. How many degrees are in a half circle?

9. If little lamb could run 8 feet in a second, how far could she run in a minute?

10. Write out in words: $7\frac{1}{5}$

The Bridge

from Chapters 1–24 to Chapter 25

fifth try

1. What is the LCM of 2, 3, 4, and 8?

2. For this problem, we'll say that the circumference of a circle is equal to $3\frac{1}{7}$ times the diameter. What would be the circumference of a circle whose diameter is equal to $2\frac{3}{4}$ miles?

3. If little lamb eats 17 ounces of food each day, how many ounces would she eat in a week?

4. Find a number z which makes $10 < z < 11$ true.

5. Change $\frac{181}{44}$ to a mixed number.

6. $6\frac{1}{6} + 7\frac{1}{3} = ?$

7. $\frac{4}{5} - \frac{3}{7} = ?$

8. Joe likes ketchup on his hamburgers. Once, when he and Darlene were having lunch together, he took a full bottle of ketchup and emptied three-fifths of it onto his burger. How much did he leave in the bottle?

Joe's
favorite food

9. Darlene ordered a 16 oz. vanilla milkshake and drank one-quarter of it. How many ounces did she drink?

10. How much of the milkshake did she not drink?

Chapter Twenty-five
Canceling

Besides the gum and the bag of bolts and nuts, Fred had many other things in his pockets. He continued to empty his pockets onto the bench. Next came out a coupon for one free ice skating lesson. It had expired a month ago. Then a baseball card with a picture of a famous nurse.

He had traded three other cards to get her.

Markie, Willy, and Julie—as Fred called them—were certainly important people. But Fred was never very comfortable carrying them around in his pocket. They were always fighting too much for Fred's taste.

In contrast, Florence Nightingale headed off to Turkey in 1854 to help those wounded in the Crimean War. The hospital she worked at was really overcrowded, the ventilation was bad, and the sewer system wasn't working. During her first winter there, 4,077 soldiers died. Less than one-tenth died because of battle wounds. More than nine-tenths died because of diseases like typhus, typhoid, and cholera. She helped clean things up.

What shall I eat? our five-and-a-half-year-old thought to himself. The last thing in his pocket was a $1\frac{1}{4}$ oz. marshmallow. Fred looked at it. He decided to eat one-tenth of it.

Your Turn to Play

1. What is one-tenth of $1\frac{1}{4}$ ounces?

2. There's an easier way to do problem 1. Interested?
It's called canceling. For some people, it's their "most favorite" part of working with fractions.

When we got to $\frac{1}{10} \times \frac{5}{4}$ the next step was to multiply top times top and bottom times bottom. Before we do that, we can save some work by doing some canceling. It's a lot like reducing fractions.

You see the 5 on top and the 10 on the bottom? Divide both by 5. Here's how it looks:

$$\frac{1}{\cancel{10}_{\,2}} \times \frac{\cancel{5}^{\,1}}{4}$$ and then we multiply and get $\frac{1}{8}$.

Your turn. Try canceling with $\frac{1}{9} \times \frac{6}{7}$

3. Here you can cancel twice $\frac{5}{6} \times \frac{4}{15}$

> General Rule #5:
>
> Canceling is only done when you're multiplying fractions.
>
> You don't do it when there is addition or subtraction.
>
> $\frac{\cancel{4} + 7}{\cancel{8}}$ No!

4. $\frac{1}{7} + \frac{1}{77} = ?$

5. $\frac{1}{7} - \frac{1}{77} = ?$

6. $3\frac{3}{5} \times \frac{5}{6} = ?$

7. 1,813 people read The KITTEN Caboodle article entitled "Fred Makes Big Mistake." One-seventh of them thought that Fred looked funny with smoke all over him. What's one-seventh of 1,813?

Chapter Twenty-six
Opposites

That $\frac{1}{8}$ oz. of marshmallow was the perfect snack for Fred. His tummy was full. He took out his handkerchief and carefully wiped his mouth. He looked at his handkerchief and there was a big black mark on it.

That's funny Fred thought to himself. That marshmallow was gray, not black. Where did all the black come from?

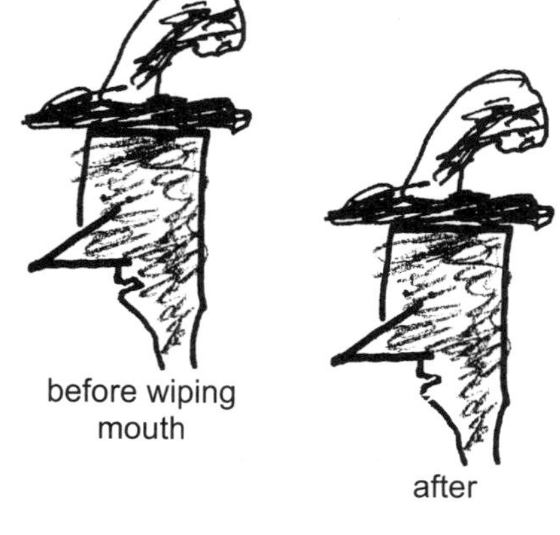

before wiping mouth

after

Then he realized that he was really dirty from the smoke of the pizza fire. What if someone saw me like this? He sat on the bench wondering what he should do.

A woman and her four-year-old daughter walked by. The girl pointed at Fred and laughed. "Look mommy! What is it?"

Her mother frowned and told her daughter, "It's not polite to point at people. He's probably just a poor homeless little four-year-old."

Fred had four thoughts that ran through his mind:

#1: I wish that they had noticed that my mouth was clean.

#2: I'm five-and-a-half-years old, not four. I'm just small for my age.

#3: I'm not homeless. I live in an office at KITTENS.

#4: I really have to get cleaned up.

Then he had the big thought: Getting cleaned up is the opposite of getting dirty. Fred knew he was onto something. His mind raced. And taking off my shoes is the opposite of putting them on. And opening my eyes is the opposite of closing them.

It's not fair that Fred should have all the fun. Now it's . . .

Your Turn to Play

1. What's the opposite of "adding six to a number"?

2. What is the opposite of "multiplying by seven"?

3. What is the opposite of "going east for an hour"?

4. A harder question: What is the opposite of "multiplying by zero"?

5. What is the opposite of changing a mixed number (like $3\frac{1}{7}$) to an improper fraction (like $\frac{22}{7}$)?

6. Is there an opposite to "take a pizza and cook it in a 450º oven for an hour"?

7. A **function** is any fixed, unchanging rule. Suppose the function I'm thinking of is "multiply by six and then add twenty-four." What is the opposite? The opposite of a function is called the **inverse function**.

8. $6\frac{2}{5} \times 6\frac{1}{4} = ?$

9. What is the inverse function to "multiply by one"?

10. There is a function called squaring which you will learn about in algebra. It means multiplying a number by itself.

The square of 5 is 25 (since 5 times itself is 25).

The square of 1,000 is 1,000,000.

What is the square of $2\frac{5}{8}$?

Complete solutions are on page 184

Chapter Twenty-seven
Area of a Rectangle

Fred thought How am I going to get cleaned up? The first thing to do would be to get rid of the chef's hat. Fred took it off and put it into the trash barrel next to the bench he had been sitting on.

with hat

But where to get washed up? He went over a checklist in his mind:

✓ My office doesn't have a bathtub or a shower.

✓ It doesn't look like it's going to rain soon.

✓ Alexander isn't home right now, so I can't use his bathroom.

a definite improvement

Then Fred saw the sign. That's the way I can get clean in a hurry. And besides, I like the alliteration.* Just a quick run through, and I'll be all clean.

Clean and Quick
Car Care
CAR WASH

Fred walked up to the man who was operating the car wash and asked, "Can I run through your car wash tunnel?"

"Why?" the man answered. "If you go in there, you'll get all wet."

"That's okay," said Fred. "I just want to get washed up."

"Listen kid. This is a *car* wash place, not a public shower. You need a car to go through there."

---

* For alliteration, the beginning of each of the words has to sound alike. The words don't have to begin with the same letter. *Quick* has the same beginning sound as *clean*.

Fred didn't know what to do. He certainly didn't have a car. Most five-and-a-half-year-olds don't own cars.

Several minutes later, the man who operated the car wash saw something very strange coming out of the car wash tunnel. It was Fred sitting on the hood of a car. He was clean. He smiled and waved at the car wash operator. He hopped off the hood and thanked the driver who had let him ride on the hood of his car. As Fred walked away, he left wet footprints on the sidewalk.

He also left a dirty spot on the hood of the car. The car wash wasn't able to wash the spot where he had been sitting. It was roughly the shape of a rectangle.

It wasn't a very large rectangle. The driver got out of the car and used his handkerchief to wipe off the dirty spot.

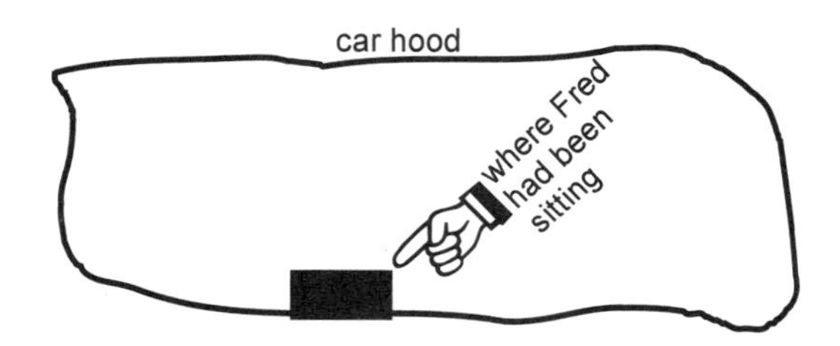

Your Turn to Play

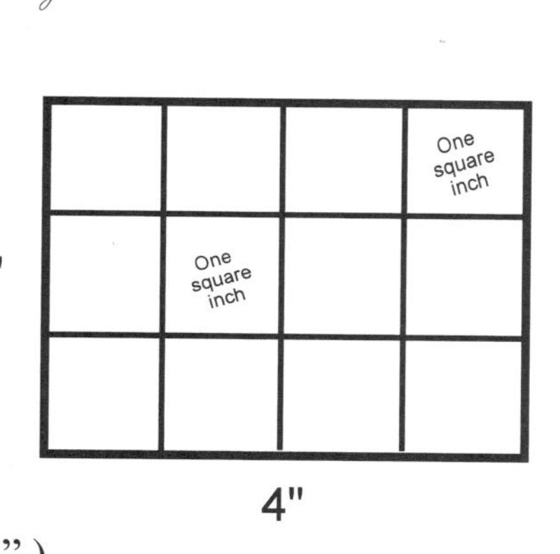

1. The rectangle measured 3" by 4". The width was three inches and the length was four inches. The area was 12 square inches.

The area of a rectangle is equal to the length times the width. (In algebra, we will write that as $A = \ell w$ where ℓ is the length and w is the width. We use a cursive ℓ, otherwise the formula would look like A = lw, which looks like "one w.")

If C. C. Coalback had been sitting on the car hood, the dirty spot would have been a rectangle that measured $9\frac{3}{4}$" by $14\frac{1}{3}$" What would have been the area of that rectangle?

2. The football field at KITTENS University is 160 feet wide. How much is that in yards? (There are three feet in a yard.)

3. The length of the KITTENS football field is 120 yards. That is the standard length for a college football field. (120 yards = 100 yards of playing field plus two 10-yard end zones.) What is the area of that field in square yards?

4. Suppose in a dream you were suddenly in the middle of a football field. Is there a way you could tell whether it was a high school or a college football field? Yes. The goal posts are different. (Did you know that?)

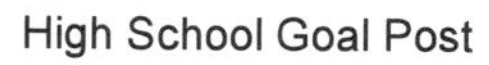

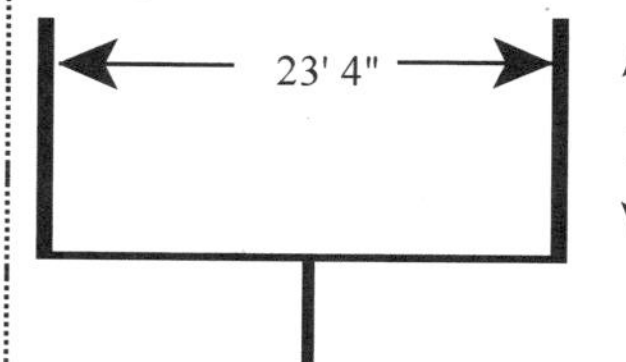

10'

Compute the areas of the two rectangles. (Convert everything to feet first.)

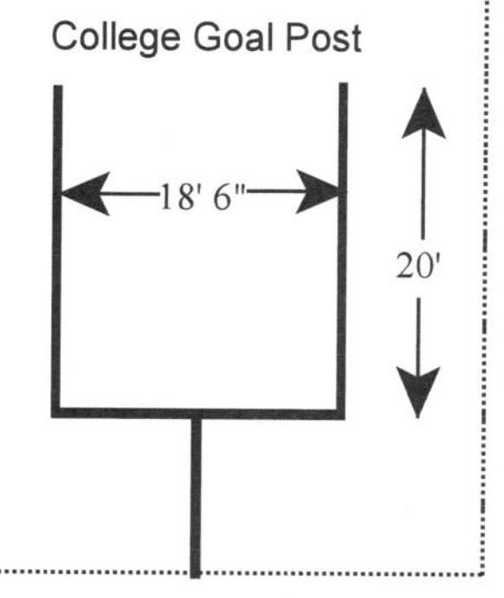

Complete solutions are on page 186

Chapter Twenty-eight
Unit Analysis

Fred had given his Florence Nightingale baseball card to the driver who had let him sit on the hood of his car. It was a way of saying, "Thank you." It is important not to forget to be polite. But Fred did forget something. He forgot the basic rule about car washes. Here it is in a box so that you don't forget it:

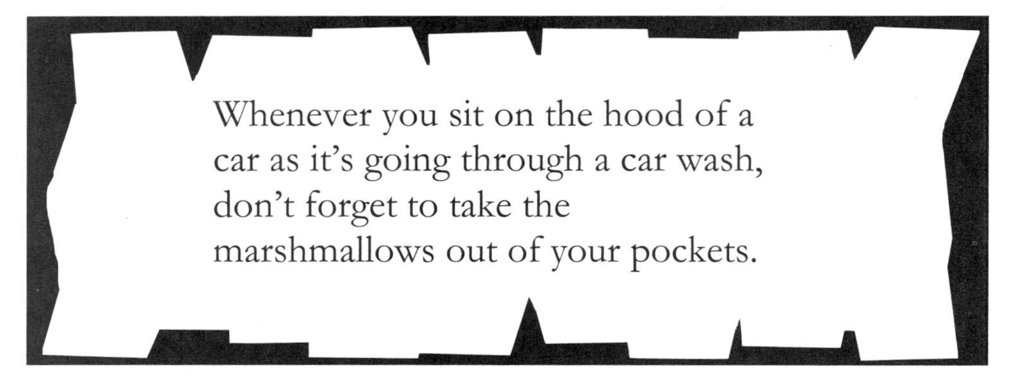

When Fred put his hand in his pocket, he realized something was wrong. It was a sticky mess. His fingers were all glued together with gray, melted marshmallow. Yuck!

He didn't know what to do. He just walked down the street with a clean face, clean clothes, and a very sticky hand. He held his hand out so that he wouldn't get the rest of himself dirty. A kitty saw Fred with his hand out and thought Fred wanted to pet her.

She purred. Fred petted her without thinking. The kitty had forgotten a basic rule:

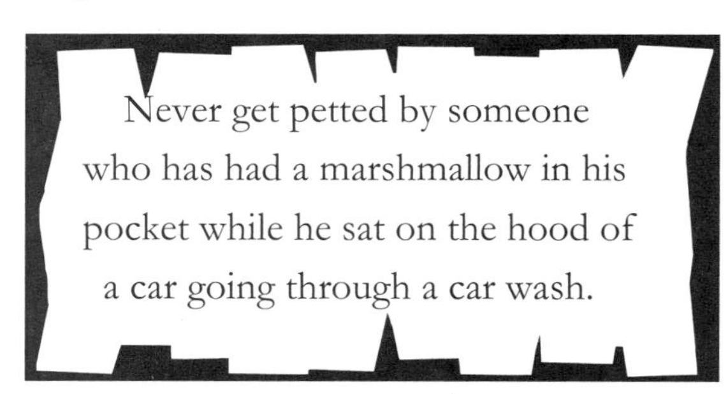

Kitty rules are sometimes more complicated than people rules.

Fred's hand was now clean. Everything seemed good. He had a bicycle. The sun was shining. His tummy was full. (Recall the marshmallow snack that Fred had had before he went through the car wash.) He was a five-and-a-half-year-old boy who felt like running down the street. So he did.

He ran at 7 feet/second for 12 seconds.

$$\frac{7 \text{ feet}}{\text{sec.}} \times \frac{12 \text{ sec.}}{1} = \frac{7 \text{ feet}}{\cancel{\text{sec.}}} \times \frac{12 \cancel{\text{sec.}}}{1} = 84 \text{ feet}$$

Wait! Stop! You canceled the . . . the . . .

They are called units.

You canceled the units! You canceled the seconds.

Yes I did.

You can't do that! The general rule you gave was "(5) Canceling is only done when you're multiplying fractions." You said that yourself.

I did.

But you're canceling units. I've never seen that before.

So you are telling me that you are in a panic because you are seeing something new?

I wouldn't exactly call it a "panic." But it is kind of new to me. Can you really do that? Can you cancel the units?

Yes, but only when you are multiplying fractions.

But you never said what happened to the kitty?

What? Did you just change the subject? Let me put the answer to your question in a footnote.✶ Let's stay on the subject, which was canceling units when you are multiplying fractions.

✶ Our gray-marshmallowed kitty was in a daze. She wandered down the street and rubbed up against the leg of a friendly little girl. The marshmallow mess was drying, and the kitty stuck to the leg of the girl. The girl went home with the kitty stuck to her leg. Her mom said, "Don't you know the rule?"

> Don't ever let a kitty brush up against your leg when the kitty has been petted by a boy who has had a marshmallow in his pocket when he rode through a car wash on the hood of a car.

Sometimes people rules are more complicated than kitty rules.

Your Turn to Play

1. If you are in a car that travels for four hours at 60 miles/hour (60 mph), how far would you go?

2. This canceling of units has a fancy name. It's part of what is called **unit analysis**. An even more fancy name is **dimensional analysis**. So if your mother calls out, "Hey, what are you doing?" you can answer, "It's okay Mom. I'm just doing some dimensional analysis." She'll think you are a genius.

Using unit analysis, here's how I change 17 miles into feet.

$$\frac{17 \text{ miles}}{1} \times \frac{5280 \text{ feet}}{1 \text{ mile}} = \frac{17 \text{ \cancel{miles}}}{1} \times \frac{5280 \text{ feet}}{1 \text{ \cancel{mile}}} = 89{,}760 \text{ feet}$$

$\frac{5280 \text{ feet}}{1 \text{ mile}}$ is called the **conversion factor**. The numerator and the denominator are equal to each other. By the third general rule,* it is equal to one. So multiplying by a conversion factor is the same as multiplying by one. And multiplying by one never hurts anything.

Your turn: change 43 gallons into quarts. (4 quarts = 1 gallon)

3. Change 888 inches into feet.

4. What is the area of a rectangle whose length is $3\frac{1}{3}$ feet and whose width is $2\frac{2}{5}$ feet?

5. What is the square of 12?

6. "Find the square of a number" is a function (a fixed rule). The opposite of "find the square of" is called "find the **square root** of." Finding the square root of a number is the inverse function of finding the square of. What is the square root of 144?

7. What is the square root of 100?

8. What is the square root of 81?

9. What is the inverse function to "take the square root of"?

Complete solutions are on page 187

---

* (1) Reduce fractions in your answers as much as possible.
(2) Fractions like $\frac{0}{4}$ are equal to 0.
(3) Fractions like $\frac{4}{4}$ (where the top and bottom are equal) are equal to 1.
(4) Division by zero doesn't make sense.
(5) Canceling is only done when you're multiplying fractions.

Chapter Twenty-nine
Subtracting Mixed Numbers

After he had run for 84 feet, Fred decided to be a little silly. He ran backwards for 84 feet. He thought that it would be a good way to illustrate inverse functions when he taught his math classes at KITTENS. He looked around. The kitty that he had petted was gone.

Then he spun around three times to the right.

Then three times to the left.

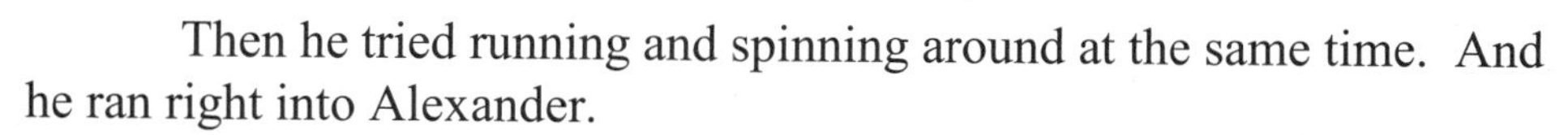

Then he tried running and spinning around at the same time. And he ran right into Alexander.

Alexander was wearing his new belt that Betty had given him. It had a beautiful eagle on the belt buckle.

Alexander is six feet tall. Fred is three feet tall.

Alexander was the first to speak. "I used to do that when I was a kid: run down the street and spin around at the same time. It was a lot of fun. I see my belt buckle made quite an impression on you. Are you okay?"

"Did you ever run into buckle belts—I mean belt buckles?" Fred asked. He was still a little dazed.

"No," said Alexander. "Mostly trees and sign posts. Once I hit a tree when I was running and spinning and carrying 16 ounces of my favorite drink—chocolate orange fizz. I used to mix up that drink myself. Some chocolate milk, some orange juice and some Sluice to make it bubbly. When I hit that tree, all I had left was $2\frac{1}{4}$ ounces. I almost cried."

Fred had trouble imagining big Alexander crying, but at some time in his life, Alexander was probably shorter than he is now. Fred thought to himself *Who would cry over losing chocolate orange fizz? That sounds pretty yucky to me.* He said to Alexander, "That was a lot of chocolate orange fizz that you lost."

Alexander nodded and said, "It seemed pretty bad at the time. Now I look back and smile. I can't think how I liked chocolate milk and orange juice and Sluice mixed together. My tastes have changed since then. I never computed how much drink I had lost."

Alexander pulled out a pocket notebook and a pen and wrote:

Started with 16 oz. Had $2\frac{1}{4}$ oz. after the spill. Want to find out how much I spilled.

Subtraction. 16 take away $2\frac{1}{4}$

$$\begin{array}{r} 16 \\ -\ \ 2\frac{1}{4} \\ \hline \end{array}$$

Wait! Stop! You can't do that. I, your reader, say that's impossible to do. You could subtract the 2 from the 16, but there is nothing to subtract the $\frac{1}{4}$ from.

Remember when you learned "borrowing" when you were doing subtraction in arithmetic?

$$\begin{array}{r} 46 \\ -\ 29 \\ \hline \end{array} \qquad \begin{array}{r} {}^{3}\not{4}{}^{1}6 \\ -\ 2\,9 \\ \hline 1\,7 \end{array}$$

Alexander borrowed . . .

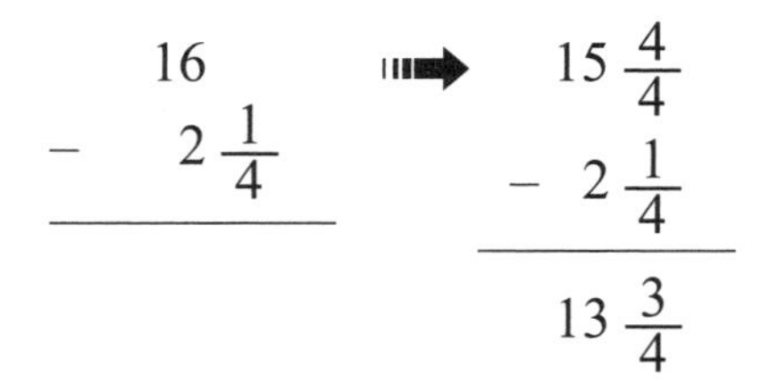

Thank you. That makes things clearer.

Your Turn to Play

1. Subtract $5\frac{1}{7}$ from 30.

2. Subtract $\frac{5}{6}$ from 200.

3. $9\frac{1}{8} - 3\frac{3}{8}$

4. $2\frac{1}{4} - 1\frac{1}{3}$

Complete solutions are on page 187

The Bridge

from Chapters 1–29 to Chapter 30

first try

Goal: Get 9 or more right and you cross the bridge.

1. $\frac{35}{44} \times \frac{2}{21} = ?$

2. If you could list 7 jobs per minute that Fred could possibly have, how many could you list in an 8-hour day?

3. If you were three times as rich as you are now, you would have \$534. How much money do you have now?

4. What is one-tenth of five and one-fifth?

5. Which is true: $\frac{3}{7} < \frac{5}{9}$ or $\frac{5}{9} < \frac{3}{7}$?

6. What is the square of $3\frac{1}{3}$?

7. Joe invited Darlene to go boating with him in his new motorboat.

When she got over to his house, she asked him where they were going with the boat. He said that the boat was already in the water and pointed to his rectangular swimming pool in the backyard. When he zoomed around in the swimming pool it made a lot of big waves. That made Joe happy. Joe's pool was $19\frac{1}{3}$ feet wide and 40 feet long. What was its area?

8. What is the square root of 81?

9. $5\frac{1}{5} - 3\frac{1}{3} = ?$

10. Fred was carrying $8\frac{1}{2}$ ounces of nuts and bolts and $\frac{1}{3}$ ounce of other stuff in his pockets. What was the total weight he was carrying in his pockets?

The Bridge

from Chapters 1–29 to Chapter 30

second try

1. Which is true: $\frac{7}{10} < \frac{5}{8}$ or $\frac{5}{8} < \frac{7}{10}$?

2. Joe had a little seed that weighed $\frac{1}{90}$ of an ounce. He planted it in his backyard and watered it every day. Each day it doubled in weight. After one day it weighed $\frac{1}{45}$ of an ounce. After two days, it weighed $\frac{2}{45}$ of an ounce. How long would it take to weigh more than an ounce?

3. $35 - 3\frac{1}{8} = ?$

4. How many tens would you have to multiply together in order to get a billion?

5. $\frac{6}{25} \times \frac{15}{24} = ?$

6. What is the inverse function to "multiply by four and then add seven"?

7. Darlene invited Joe over to make cookies. Joe's cookies were larger than Darlene's.

The weight of 2 of Joe's cookies equaled 9 of Darlene's. Using a conversion factor, compute how many of Joe's cookies would weigh as much as 20 of Darlene's.

8. Express 667 in Roman numerals.

9. If you had 67 bolts, how many more bolts would you need to have a million bolts?

10. If we share six of Joe's cookies equally among eight people, how much does each person receive?

The Bridge
from Chapters 1–29 to Chapter 30

third try

---

1. Reduce as much as possible $\frac{35}{42}$

2. What would you multiply $63 by in order to get $63,000?

3. Joe was driving his motorboat around in his swimming pool. When he drove it into the diving board, it made a neat rectangular hole in his boat. The hole was $2\frac{3}{4}$ inches by $8\frac{1}{8}$ inches. What was the area of that rectangle?

4. $7\frac{1}{7} - 2\frac{5}{7} = ?$

5. If you take two whole numbers and add them together, will you always get an answer that is an integer?

6. 739 + 28 + 38 + 14 = ?

7. Twelve minutes is what fraction of an hour? Reduce your answer as much as possible.

8. What is one-seventh of eight and two-fifths?

9. What is the inverse function to "divide by $\frac{3}{7}$ "?

10. Darlene made some popcorn for Joe and herself. For every 2 pieces of popcorn that Darlene ate, Joe ate 19. Using a conversion factor, compute how many pieces Joe ate, if Darlene ate 14 pieces.

The Bridge

from Chapters 1–29 to Chapter 30

fourth try

1. Compare $\frac{3}{4}$ and $\frac{2}{3}$

2. If you had a million dollars, how many more dollars would you need to have two billion dollars?

3. $45\frac{1}{5} - 20 = ?$

4. Reduce as much as possible $\frac{28}{42}$

5. If Fred makes \$500 each month working at KITTENS, how much would he make in ten years?

6. $\frac{3}{7} = \frac{?}{35}$

7. Write 539 in Roman numerals.

8. $\frac{27}{40} \times \frac{8}{9} = ?$

9. Stanthony could wipe off a table in 45 seconds. How long would it take him to wipe off 87 tables? Express your answer in minutes and seconds. (Use a conversion factor.)

10. If Betty were seven times as tall as she is, she would be 476 inches tall. How tall is she? Give your answer in feet and inches.

The Bridge

from Chapters 1–29 to Chapter 30

fifth try

1. $\frac{16}{21} \times \frac{7}{64} = ?$

2. CVI + XC = ? (Express your answer in Roman numerals.)

3. What is one-third of twelve and three-fourths?

4. What is the square of $4\frac{1}{4}$?

5. If Fred could run 77 feet in a minute, how far could he run in 5 minutes?

6. $14\frac{1}{6} - 5\frac{5}{6} = ?$

7. Compare $\frac{7}{8}$ and $\frac{57}{64}$

8. If you had a pepperoni pizza with a diameter of 17", what would be its radius?

9. Is there a cardinal number x such that $17 < x < 18$? If so, give an example. If not, explain why there isn't.

10. Joe's boat had a rectangular windshield. Its area was 360 square inches. It was 10 inches high. How wide was it?

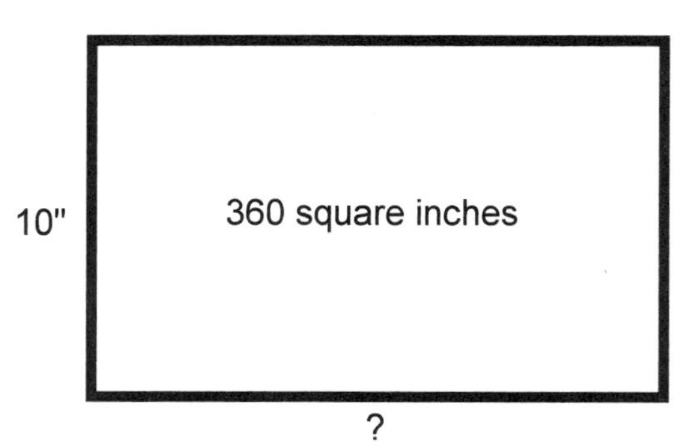

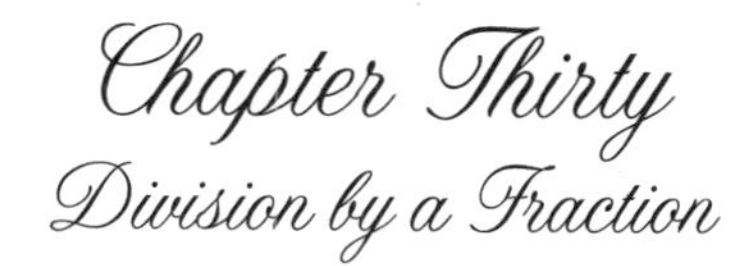

Chapter Thirty
Division by a Fraction

They stood there a moment thinking about the chocolate orange fizz that Alexander had spilled years ago. Then Fred commented to Alexander what a mess it would have been if he had been riding a bike when he hit the tree.

"I would have spilled the whole thing—all sixteen ounces," Alexander replied. "But at least I wouldn't have cracked my head on the tree. I always wear my helmet when I'm bicycling. By the way, did you get a helmet when you bought your bike?"

"Oops. I knew I forgot something," said Fred. "In my excitement I forgot about getting a helmet. I'll need one before I can ride my new bike."

"But when you and I and Betty were at PieOne, you told us that you were out of money, and that you had spent it all on a new bike," Alexander said. "How are you going to buy a helmet?"

Fred shrugged his shoulders. "I don't know."

They were both quiet for a moment. Fred rubbed his forehead* and said, "I guess I'll just go back to my office and look at my new bike. I haven't taken it out of the box yet. Maybe they packed a free helmet in the box."

They parted. Fred walked (without spinning) back to his office. He climbed the stairs up to the third floor. He mumbled something about ordinal numbers and went into his office.

Books lined the office walls. Fred had arranged them alphabetically. At one end was *Abalone—How to Cook Them,* followed by *Abdominal Muscles—Great Ways to Make Your Tummy Look Cool.* At the other end was *Zoolatry Is a Mistake.*** In the middle of his office was his desk and the bicycle box. Fred's mail had been left on top of the box.

---

* The preferred way to say that word is FOUR-id, not FOUR-head. Look in any good dictionary.

** Zoolatry (zoe-ALL-a-tree) means worshiping animals or paying excessive attention to them.

Fred often got a lot of mail. Much of it was because of his teaching at KITTENS University. Everyone was interested in the "Fred Way" of teaching mathematics. Some teachers wrote to him, asking him for help in teaching. Universities wrote to him, offering him teaching jobs. Students he had taught in previous years wrote to him to thank him for the wonderful memories they had of his teaching.

Of course, there was some junk mail.

Fred threw it in the trash.

Fred had 36 letters to read. If he read 3 letters per minute, it would take him 12 minutes to get through his mail. $36 \div 3 = 12$.

If he read 4 letters per minute, it would take him 9 minutes to get through his mail. $36 \div 4 = 9$.

Actually, he could read $4\frac{4}{5}$ letters each minute. So dividing $4\frac{4}{5}$ into 36 would tell us how long it would take Fred to get through his mail.

$$36 \div 4\frac{4}{5} \text{ which is } \frac{36}{1} \div \frac{24}{5}$$

First, let me tell you **how** to divide by fractions. Then I'll tell you **why** the rule is true.

The rule: If you want to divide by $\frac{24}{5}$

just multiply by $\frac{5}{24}$ instead.

$\frac{5}{24}$ is called the **reciprocal** of $\frac{24}{5}$ (re-SIP-re-kull)

So, to divide by a fraction, just multiply by its reciprocal.

Your Turn to Play

1. $4\frac{2}{5} \div 3\frac{1}{3} = ?$

.......COMPLETE SOLUTION.......

1. $4\frac{2}{5} \div 3\frac{1}{3} = \frac{22}{5} \div \frac{10}{3} = \frac{22}{5} \times \frac{3}{10} = \frac{\overset{11}{\cancel{22}}}{5} \times \frac{3}{\underset{5}{\cancel{10}}}$

$= \frac{33}{25} = 1\frac{8}{25}$

(The chapter is not over. Keep reading.)

Hey! That was easy. When you want to divide by a fraction, you just turn it upside down and multiply instead.

Or, as they sometimes say, **invert** and multiply.

Now stop here for a second. I, as your loyal reader, need to do something. I'll be right back.

[46 MINUTES PASS]

Thanks for waiting. I went and checked all the other fractions books that I could find. None of them told me WHY you take the reciprocal and multiply when you want to divide by a fraction. How come you know?

It was in one of Fred's math lectures. I just paid attention.

Okay. Shoot. Tell me why it's true: "To divide by a fraction, just multiply by its reciprocal."

Here's what I heard in Fred's class:

Fred wrote on the board: Multiplying by $\frac{3}{2}$ is a function.

Everyone nodded yes. It was a fixed rule.

Then Fred asked Joe, "What is the inverse function to "multiplying by $\frac{3}{2}$" ?

Joe asked, "You mean what's the opposite of multiplying by $\frac{3}{2}$?"

"Yes," said Fred.

"That's a cinch. The opposite of multiplying by $\frac{3}{2}$ is dividing by $\frac{3}{2}$ because you can undo multiplying by any number by just dividing by that number."*

Fred drew a chart on the board:

the function	its inverse
$\times\frac{3}{2}$	$\div\frac{3}{2}$

Then Fred turned to Darlene (who had been talking on her cell phone) and asked her the same question that he had asked Joe.

✶ Joe was almost right. He should have said, "You can undo multiplying by any *non-zero* number by just dividing by that number."

"Darlene, what is the inverse function to multiplying by $\frac{3}{2}$?"

She hadn't heard Joe's answer and gave a different answer. She said, "If I multiplied something by $\frac{3}{2}$ I could get back to the original number by multiplying by $\frac{2}{3}$."

"That's right," Fred said.

Joe stood up. "I thought you said my answer was right."

"You're both right," Fred said.

Darlene pulled on Joe's sleeve, urging him to sit down.

Fred enlarged his chart on the board:

the function	its inverse
$\times \frac{3}{2}$	$\div \frac{3}{2}$ [Joe's answer]
$\times \frac{3}{2}$	$\times \frac{2}{3}$ [Darlene's answer]

So $\div \frac{3}{2}$ and $\times \frac{2}{3}$ have the same effect—they both undo $\times \frac{3}{2}$

If you want to divide by $\frac{3}{2}$ you can just multiply by $\frac{2}{3}$

the function	its inverse
$\times \frac{3}{2}$	$\div \frac{3}{2}$ } the same!
$\times \frac{3}{2}$	$\times \frac{2}{3}$ }

This was a perfect example of the use of inverse functions. About 50% (one-half) of the students understood the proof when Fred presented it. The other half (like me) had to go home and play with it a while before it became clear.

Your Turn to Play

1. $10\frac{1}{10} \div 6\frac{1}{5} = ?$

2. $\frac{1}{6} \div \frac{1}{4} = ?$

3. Fred had 36 letters to read. He could read $4\frac{4}{5}$ letters each minute. How long would it take for him to get through his mail?

4. Now, do the previous problem showing the units (letters and minutes) in each step. This will show that division was the correct thing to do.

5. $8\frac{1}{3} - 7\frac{2}{3} = ?$

6. What is the square root of 49?

7. What is the area of a rectangle whose length is 12' 3" and whose width is 4' 4"?

8. Name an integer that is not a whole number.

9. What is the least common multiple (LCM) of 10 and 12?

10. $\frac{1}{2} \div \frac{1}{4} = ?$

Complete solutions are on page 188

So in $7\frac{1}{2}$ minutes Fred had worked through the 36 letters he had received. He could now work on opening the bicycle box and see whether there was a free helmet packed inside.

How to open the box? Earlier in the day he had tried cutting the tape with a knife. That was a disaster.

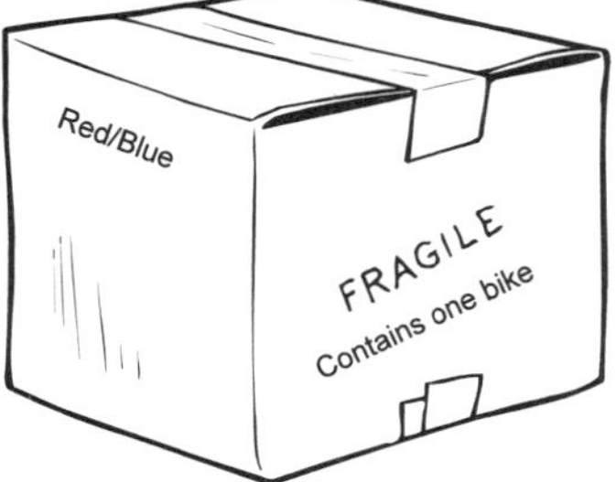

What to do? He looked around his office. He spotted the perfect book on one of his shelves: It was Prof. Eldwood's *Box Opening the Modern Way*, 1852. The book was right there between *Bow-wow, the Autobiography of* and *Boyle's Law.*✶

Professor Eldwood had many suggestions in his *Box Opening* book:

1. Try cutting the tape with an 18", 13-lb. knife.
2. Throw the box out of the window and see if it breaks open.
3. Put some candy inside the box, and then let some mice chew their way into the box.

✶ Boyle's law talks about pressure and volume of a gas. Suppose you have a rubber frog that is filled with air and you are stepping on it lightly. Boyle's law says that if you multiply the pressure by two (step on it twice as hard), you will divide the volume of the frog by two. If you get some friends and you all step on the frog and multiply the pressure by seven, then the volume of the frog will be divided by seven. And if you multiply the pressure by $2\frac{1}{3}$ you will divide the frog's volume by $2\frac{1}{3}$.

To divide the frog's volume by $2\frac{1}{3}$ means $\div 2\frac{1}{3}$ which means $\div \frac{7}{3}$ which means $\times \frac{3}{7}$ which means the frog will be three-sevenths of its original size.

Fred looked at these suggestions. None of them seemed very good. If he threw the box out of the window, it would hurt the bike. If he could put candy inside the box, he wouldn't need any mice. The box would already be open. Fred continued reading:

4. Tape a stick of dynamite to the side of the box.
5. Get a hose and wet the box down. When the cardboard becomes soft and mushy, the box can be opened easily.
6. Peel the tape off.

That's it! Fred thought. Professor Eldwood is such a smart man.

Fred's little fingers went to work on the tape. Off came the first piece. It came off so easily. It must be rotten tape. He held it up and looked at it.

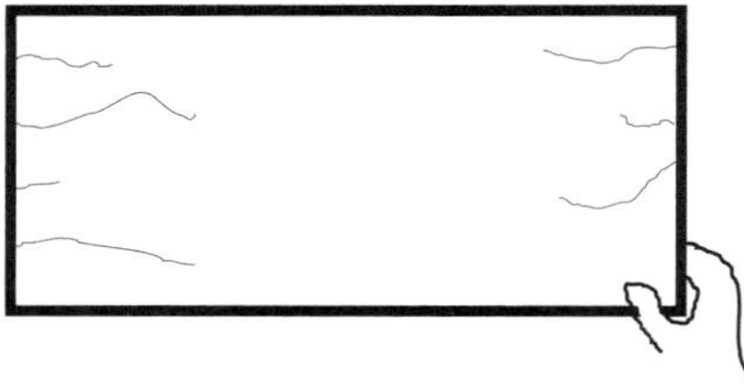

It's a pretty rectangle. All the corners are square.* He laid the tape on his desktop. With a pencil and a ruler he drew one of the **diagonals** of the rectangle.

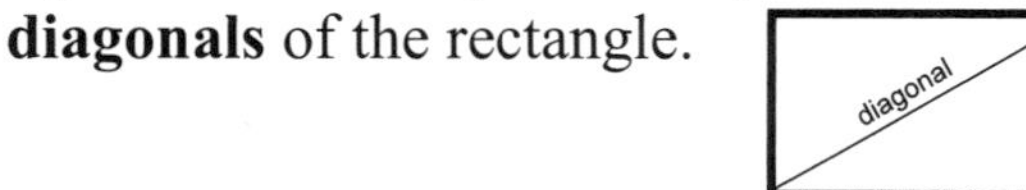

Then he marked one vertex with a little square. In geometry, that is the standard way of saying, "This is a right angle."

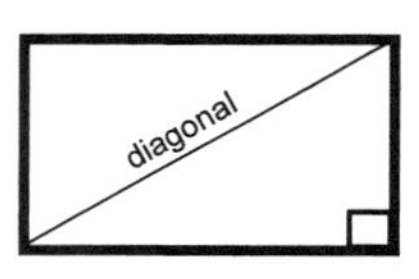

He got out a blank piece of paper and began to play.

---

* When you get to geometry, each corner is called a **vertex**.

A "square corner" is called a **right angle**.

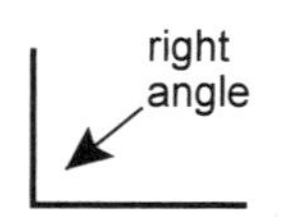

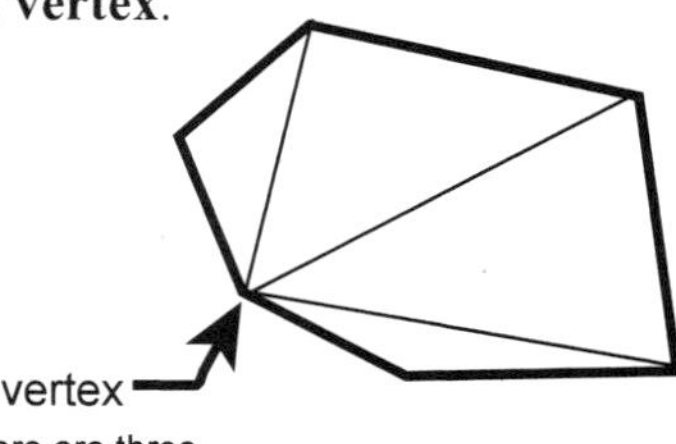

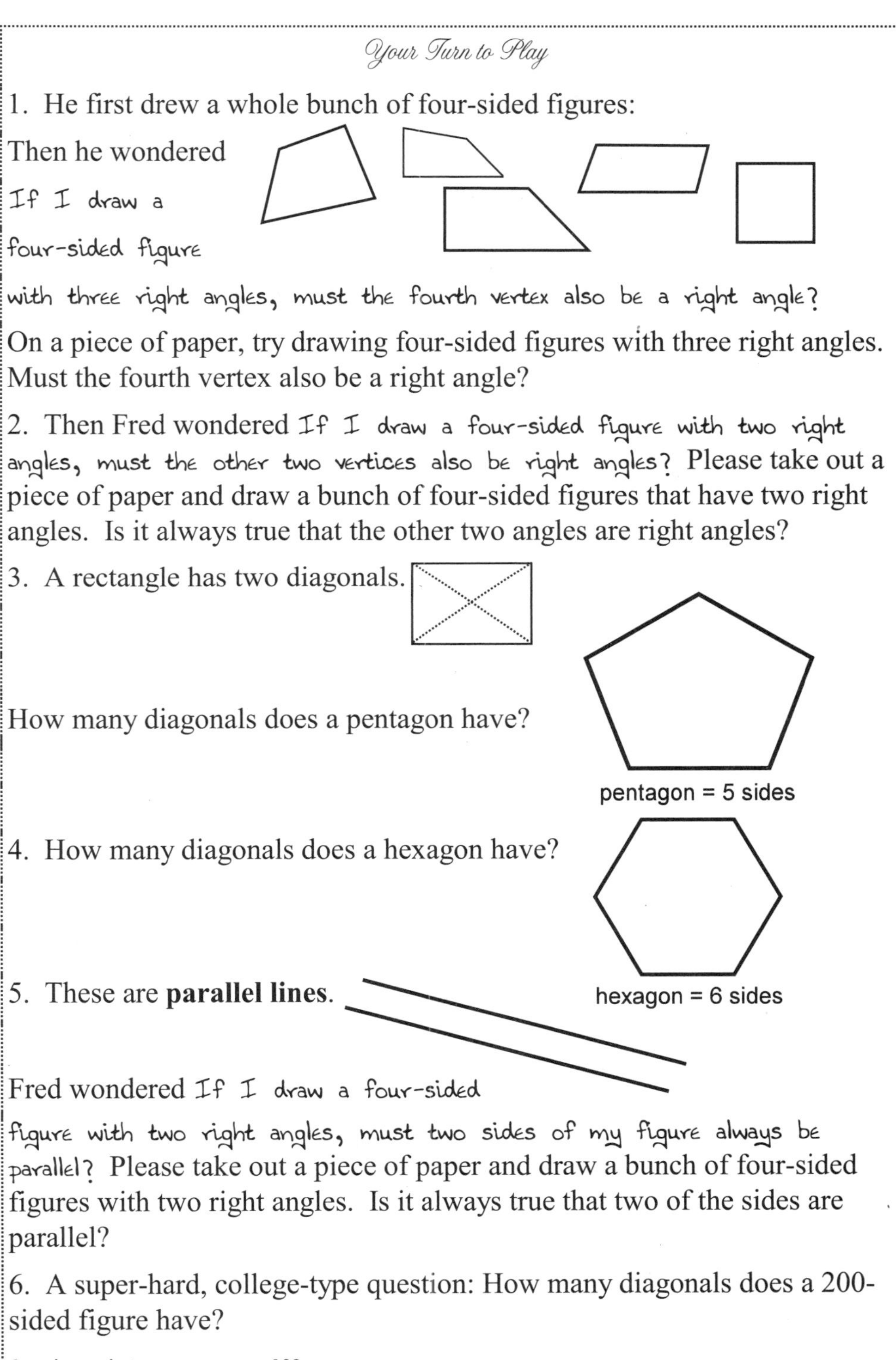

Your Turn to Play

1. He first drew a whole bunch of four-sided figures:

Then he wondered If I draw a four-sided figure with three right angles, must the fourth vertex also be a right angle?

On a piece of paper, try drawing four-sided figures with three right angles. Must the fourth vertex also be a right angle?

2. Then Fred wondered If I draw a four-sided figure with two right angles, must the other two vertices also be right angles? Please take out a piece of paper and draw a bunch of four-sided figures that have two right angles. Is it always true that the other two angles are right angles?

3. A rectangle has two diagonals.

How many diagonals does a pentagon have?

pentagon = 5 sides

4. How many diagonals does a hexagon have?

hexagon = 6 sides

5. These are **parallel lines**.

Fred wondered If I draw a four-sided figure with two right angles, must two sides of my figure always be parallel? Please take out a piece of paper and draw a bunch of four-sided figures with two right angles. Is it always true that two of the sides are parallel?

6. A super-hard, college-type question: How many diagonals does a 200-sided figure have?

Complete solutions are on page 189

Chapter Thirty-two
Estimating Answers

Fred got the last piece of rotten tape off the top of the box. His heart was beating very fast. His new bike would let him get to classes faster. It would make him taller when he was on it. It would tell the world that he was no longer a baby. With that last piece of tape off, the whole box fell open.

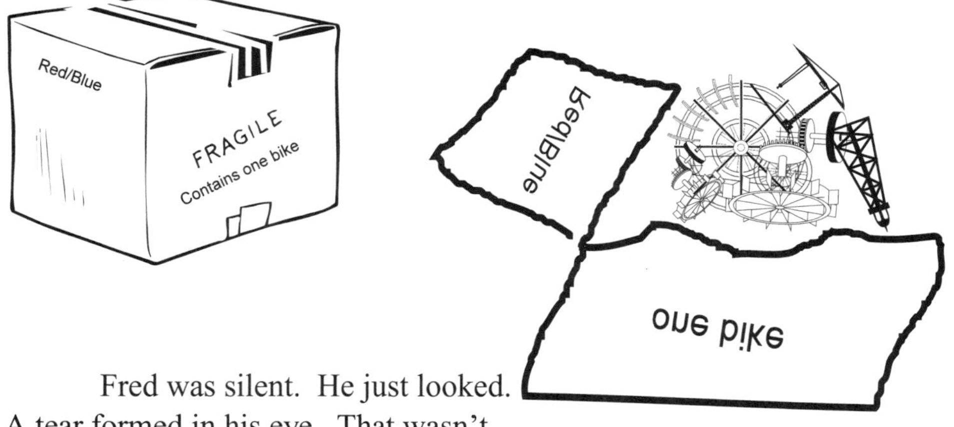

Fred was silent. He just looked. A tear formed in his eye. That wasn't a bike. That wasn't even the parts for a bike.

It was junk.

He felt cold. He grabbed a blanket and went and sat in a corner of his office. And cried.

Your Turn to Play

1. When you're upset, it's often hard to think straight. In the box, there were 11 gears, 12 rods, 9 wires, and 10 motors. Without adding those up, would you say that there were around 40 pieces or 400 pieces or 4000 pieces?

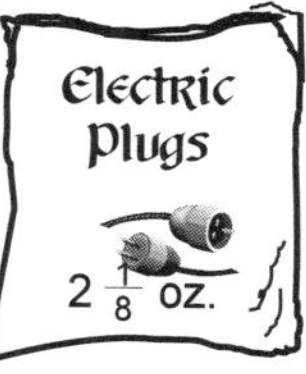

2. There were six bags of plugs. Each bag looked like →
Without multiplying it out, would the six bags weigh closer to 6 oz. or 12 oz. or 24 oz.?

3. As Fred sat in the corner, something in his pocket hurt him. He pulled out the $8\frac{1}{2}$ oz. bag of nuts and bolts that he had been carrying around. He threw them across the room onto the pile of junk.

He stood up and walked over to the mess. He almost stepped on a very unusual remote control. (Have you ever seen one this large?) It had about 42 rows of buttons and 4 buttons in each row. *That's a lot of buttons* Fred thought to himself. Estimate how many buttons without multiplying it out. 80 buttons or 200 buttons or 400 buttons?

4. Just by estimating, you can sometimes tell when something is wrong. Find the two that are wrong. Do this just by looking.

A) $3\frac{1}{3} \times 3\frac{1}{4} = 8\frac{1}{12}$

B) $6\frac{5}{6} \times 2\frac{1}{3} = 15\frac{17}{18}$

C) $6\frac{2}{5} + 7\frac{1}{7} = 13\frac{19}{35}$

D) $9\frac{3}{4} \times 9\frac{5}{8} = 100\frac{7}{8}$

Complete solutions are on page 190

Gears, wires, rods, motors, electrical plugs, and a remote control Fred thought. *What can I do with these? I know! I will build a robot.*

The Final Bridge

All 32 Chapters

first try

Goal: Get 13 or more right and you finish *Life of Fred: Fractions*

1. What is the area of a sheet of paper that measures $8\frac{1}{2}$ inches by 11 inches?

2. $4\frac{1}{7} \times 7\frac{1}{4} = ?$

3. Is it possible to draw a five-sided figure with three right angles?

4. $9\frac{1}{8} \div 4\frac{1}{4} = ?$

5. If seven bolts weigh four ounces, how much would eleven bolts weigh? (Hint: use a conversion factor.)

6. Change $7\frac{1}{8}$ into an improper fraction.

7. To change a diameter of a circle into its circumference, you use the function "multiply by $3\frac{1}{7}$" What would be the diameter of a circle whose circumference is 11"? (Hint: Use the inverse function.)

8. Find the sum of six hundredths and four hundredths. Express your answer as a fraction that is reduced as much as possible.

9. Which is smaller: $\frac{4}{9}$ or $\frac{43}{100}$?

10. Find the LCM of 2 and 40.

11. One fifth of an hour is how many minutes?

12. What is the square of $3\frac{7}{8}$?

13. How many lines of symmetry does this arrow have?

14. Subtract $20\frac{7}{8}$ from $40\frac{1}{4}$

15. If little lamb jumps up 7 times each minute, how many times does she jump up in an hour?

The Final Bridge

All 32 Chapters

second try

---

1. Change $\frac{151}{17}$ into a mixed number.

2. Fred spent $\frac{3}{4}$ of an hour working on a math problem. While he was working on that problem, he spent $\frac{2}{3}$ of an hour holding a pencil. How long did he spend on the math problem while not holding a pencil?

3. $3\frac{1}{10} \times 10\frac{1}{4} = ?$

4. There were 12 rods in the bike box. Each weighed $3\frac{5}{8}$ ounces. How much did all 12 rods weigh?

5. $27\frac{3}{4} + 88\frac{3}{4} = ?$

6. How much money is 3 pennies plus 4 nickels plus 30 dimes?

7. MMMXXXIII ÷ IX = ? Express your answer in Roman numerals.

8. What is one-seventh of 397?

9. How many 7's do you have to add up to get 539?

10. Express 9 as an improper fraction.

11. What is the square root of 64?

12. $4\frac{1}{2} \div 2\frac{3}{4} = ?$

13. What is the reciprocal of $\frac{25}{3}$?

14. Subtract 39 from $78\frac{48}{67}$

15. Change 6 miles into feet.

The Final Bridge

All 32 Chapters

third try

1. $\frac{1}{10} \div 2\frac{1}{10} = ?$

2. There are four quarts in a gallon. How many gallons would $5\frac{1}{4}$ quarts be?

3. $6\frac{1}{6} + 3\frac{2}{3} + 4\frac{1}{2} = ?$

4. How many lines of symmetry does this arrow have?

5. What is the LCM of 3 and 40? (Least Common Multiple)

6. Find one-sixth of $5\frac{1}{6}$

7. Give an example to show that division is not commutative.

8. What is the area of a square that is 5 miles on each side?

9. Subtract $\frac{7}{8}$ from a billion.

10. The 10 motors in the bike box weighed $52\frac{1}{7}$ pounds. If all the motors were identical, how much did each of them weigh?

11. $5\frac{1}{7} \times 8\frac{1}{7} = ?$

12. $\frac{3}{7} + \frac{43}{100} = ?$

13. Which is larger: $\frac{3}{7}$ or $\frac{43}{100}$?

14. How many diagonals does this figure have?

15. If Betty were five times richer, she would have $18,700. How much money does she have?

The Final Bridge

All 32 Chapters

fourth try

1. What is one-fifth of $25\frac{1}{5}$?

2. One-twelfth of an hour is how many minutes?

3. $6\frac{1}{3} \times 2\frac{1}{6} = ?$

4. Can a triangle have three right angles? Can it have two right angles? Can it have one right angle?

5. $\frac{1}{6} - \frac{1}{12} = ?$

6. What is the area of a rectangle that measures $2\frac{5}{8}$ feet by $4\frac{5}{8}$ feet?

7. XCIX × IX = ? Express your answer in Roman numerals.

8. $\frac{1}{6} \div \frac{1}{12} = ?$

9. There are four quarts in a gallon. Convert $5\frac{1}{4}$ gallons into quarts.

10. Just by estimating, find the two that are wrong.

    A) $6\frac{1}{6} \times 8\frac{5}{8} = 53\frac{9}{48}$

    B) $20\frac{5}{8} + 10\frac{1}{7} = 30\frac{43}{56}$

    C) $7\frac{3}{4} + 7\frac{1}{13} = 13\frac{43}{52}$

    D) $5\frac{1}{5} \times 3\frac{3}{5} = 24\frac{3}{25}$

11. What is the inverse function to "take the square root of"?

12. $\frac{2}{3} + 2\frac{2}{3} + \frac{1}{4} = ?$

13. Reduce as much as possible $\frac{35}{63}$

14. Change $37\frac{1}{8}$ into an improper fraction.

15. If Joe ate $\frac{13}{16}$ of a pizza, how much of the pizza did he leave uneaten?

The Final Bridge

All 32 Chapters

fifth try

1. What is the square root of 100?

2. Change $\frac{671}{19}$ into a mixed number.

3. If Fred were to build a robot that is one forty-fourth of a furlong tall, how tall would that be in yards? (A furlong is 220 yards.)

4. $16\frac{3}{4} + 27\frac{3}{8} = ?$

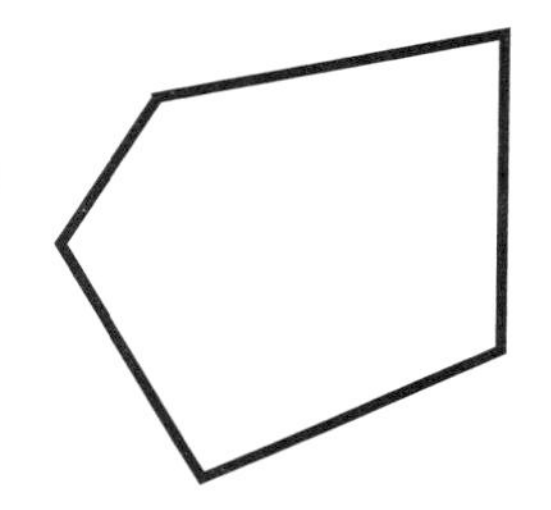

5. How many diagonals does this figure have?

6. $6\frac{1}{8} - 4\frac{1}{4} = ?$

7. Reduce as much as possible $\frac{40}{88}$

8. $33\frac{1}{3} \times \frac{1}{100} = ?$

9. Which is smaller: $\frac{3}{5}$ or $\frac{61}{100}$?

10. $3 + 4 = 4 + 3$ and $7\frac{1}{6} + 6\frac{1}{6} = 6\frac{1}{6} + 7\frac{1}{6}$ are two examples of what law of addition?

11. What is the square of 17?

12. $3\frac{1}{4} \div 3\frac{1}{4} = ?$

13. $3{,}498 \times 9\frac{1}{8} \times 0 = ?$

14. How many lines of symmetry does this figure have? (Careful! The answer isn't two.)

15. One eighth of a week is how many hours?

The Bridge
answers

from p. 32 — *first try*

1. He would make 12 times as much, which is $7,200 a year.
2. The diameter is twice the radius. Twice 53" is 106".
3. 233 – 17 = 216 feet.
4. T, T, F.
5. Four million, three hundred forty-five thousand, two hundred eleven.
6. 893 + 237 = 1130.
7. There are 60 minutes in one hour. In 24 hours, there are 24 × 60 = 1440 minutes.
8. 93rd is an ordinal number.
9. To convert inches to feet, divide by 12. 200" ÷ 12 = $12\overline{)200}$ = 16 R 8 = 16' 8".
10. 1,000,000,000 – 1,000,000 = $999,000,000 (or nine hundred ninety-nine million dollars).

from p. 33 — *second try*

1. It will be any number less than 55. It could be 54 or 2 or 0 or 54.99 or 54½.
2. 3,864,999.
3. A diameter is twice the radius. Twice 39" is 78".
4. You divide by 12. 38" = 3' 2".
5. 593 + 498 = 1091.
6. 449 ÷ 13 = 34 R 7 (or 34.538 or $34\frac{7}{13}$).
7. y is to be a number between 7 and 10. It could be 8 or 9 or 7½ or 9.99. (Not 7 or 10).
8. 302 – 18 = $284.
9. One-third of 180 = $\frac{1}{3} \times 180 = 3\overline{)180}$ = 60
10. Four billion, seventeen.

from p. 34 — *third try*

1. Eighteenth (or 18th).
2. *Third* is not a cardinal number. It is an ordinal number.
3. 400 + 3 + 38 + 5 = $446.
4. To convert inches to feet, divide by 12. 300 ÷ 12 = 25 feet
5. F, T, F.
6. 1,000,000,000 – 3,000,000 = $997,000,000 (or nine hundred ninety-seven million dollars).
7. A radius is half of a diameter. Half of 88' is 44' (or forty-four feet).
8. $800 × 12 = $9,600.
9. 7,000,000,366.
10. 963 + 877 = 1840.

The Bridge
answers

from p. 35 — *fourth try*

1. Two billion, six hundred thousand.
2. 70.
3. 1590.
4. One million, one. (Or 1,000,001.)
5. A diameter is twice the radius. $2 \times 8'' = 16''$.
6. w is to be a number between 50 and 100. It could be 51 or 63 or 99 or $99\frac{1}{2}$. (But not 50 or 100.)
7. $88'' = 7'\ 4''$.
8. T, T, T.
9. $635.
10. $60 \times 39 = 2340$ minutes.

from p. 36 — *fifth try*

1. $3{,}000{,}007 - 1{,}000{,}000 = \$2{,}000{,}007$.
2. This will be any number greater than 10. For example: 11 or 39 or $10\frac{1}{2}$ or 5558939.
3. 8,288,106.
4. $12 \times \$900 = \$10{,}800$.
5. One-third of 240 = $\frac{1}{3} \times 240 = 3\overline{)240}$ = 80
6. Yes, *fifth* is an ordinal number.
7. A radius is half of a diameter. One-half of 50 = $\frac{1}{2} \times 50 = 25$ miles.
8. Two million, eight hundred twenty-seven.
9. $82 - 59 = 23$ inches. (Did you remember to write "inches"?)
10. $7\overline{)388}$ = 55 R 3 or $55\frac{3}{7}$

$$\begin{array}{r} 55\ \text{R}\ 3 \\ 7\overline{)388} \\ -35 \\ \hline 38 \\ -35 \\ \hline 3 \end{array}$$

from p. 52 — *first try*

1. First double: 2. Second double: 4. Third: 8. Fourth: 16. Fifth: 32. Sixth: 64. Seventh: 128. Eighth: 256. Ninth: 512. Tenth: 1024. Ten doubles.
2. $783 + 697 = 1480$.
3. (see diagram →)

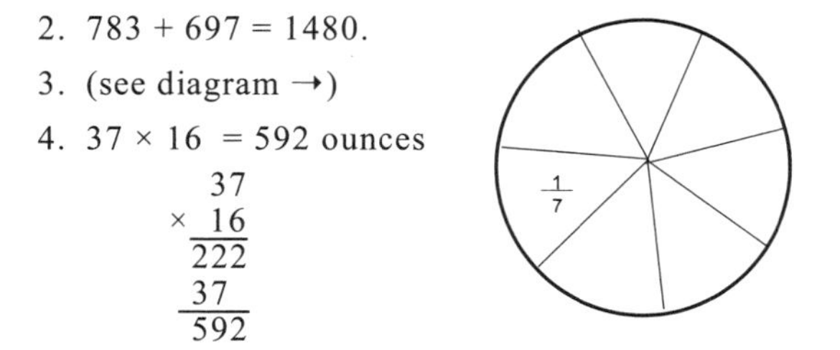

4. $37 \times 16 = 592$ ounces

$$\begin{array}{r} 37 \\ \times\ 16 \\ \hline 222 \\ 37 \\ \hline 592 \end{array}$$

The Bridge
answers

5. 7 minutes × 60 seconds/minute = 420 seconds.

6. $\frac{7}{60}$

7. If the sign read *You must be taller than 42"* then $x > 42$ would have been correct. Instead, the sign indicated that you could either be 42" or taller, so the correct answer is $x \geq 42$.

8. $\frac{1}{8} < \frac{1}{6}$

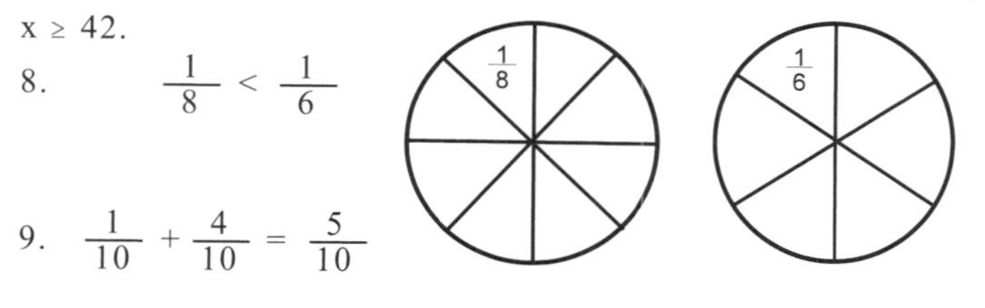

9. $\frac{1}{10} + \frac{4}{10} = \frac{5}{10}$

and reducing $\frac{5}{10}$ (by dividing top and bottom by 5) we get $\frac{1}{2}$

10. Nine hundred ninety-nine million, nine hundred ninety-nine thousand, nine hundred ninety-nine dollars.

$$\begin{array}{r} 1{,}000{,}000{,}000 \\ - \quad 1 \\ \hline \$\ 999{,}999{,}999 \end{array}$$

from p. 53 — *second try*

1. *First* is the smallest ordinal. If there were an ordinal smaller than *first*, then *first* wouldn't be *first*, would it?

2. Zero times anything is always zero.

3.

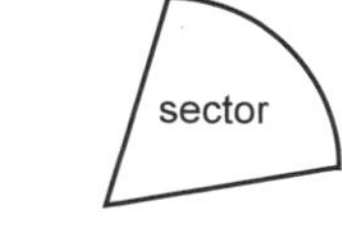

4. If I had asked, "How much bigger is 8 than 3?" you would have said, "5." You would have subtracted: 8 – 3 = 5. To find out how much taller 29,028' is than 53', you subtract.

$$\begin{array}{r} 29028 \\ - \quad 53 \\ \hline 28975 \end{array}$$

The mountain would be 28,975 feet taller than Fred.

5. $\frac{10}{60}$ I reduce this fraction by dividing top and bottom by 10 to get $\frac{1}{6}$

I could have reduced the fraction in two steps instead of one.
First, I could have divided by 2: $\frac{10}{60} = \frac{5}{30}$

Then, I could divide by 5: $\frac{5}{30} = \frac{1}{6}$

You get the same answer either way.

6. 121" = 10' 1"

$$\begin{array}{r} 10\ R\ 1 \\ 12\overline{)\,121} \\ -12 \\ \hline 1 \end{array}$$

The Bridge
answers

7. 42973 + 558 = 43,531

8. \$7/minute × 60 minutes/hour = \$420/hour

\$420/hour × 24 hours/day = \$10,080/day

\$10,080/day × 7 days/week = \$70,560 saved each week

9. $\frac{4}{9} + \frac{2}{9} = \frac{6}{9}$ which reduces to $\frac{2}{3}$

10. There are many ways to reduce $\frac{36}{144}$

Way #1: The fast way—in one step. Divide top and bottom by 36: $\frac{36}{144} = \frac{1}{4}$

Most people don't use this way since they don't know that 36 divides evenly into 144.

Way #2: Using three steps. Divide top and bottom by 2: $\frac{36}{144} = \frac{18}{72}$

Then divide top and bottom by 9: $\frac{18}{72} = \frac{2}{8}$

Then divide top and bottom by 2: $\frac{2}{8} = \frac{1}{4}$

Way #3: Using four steps. Divide by 2; divide by 2; divide by 3; divide by 3.

$$\frac{36}{144} = \frac{18}{72} = \frac{9}{36} = \frac{3}{12} = \frac{1}{4}$$

Moral: The better you are at arithmetic, the fewer steps you have to take.

from p. 54 — *third try*

1. $\frac{8}{60} = \frac{2}{15}$ I divided top and bottom by 4.

2. Three-eighths is larger than one-eighth. (It's three times larger.)

3. If you are a millionaire, then either you have a million dollars or you have more than a million dollars. $y \geq \$1,000,000$.

4. 10 × 10 × 10 × 10 × 10 × 10 × 10 × 10 × 10 = 1,000,000,000.

You need nine 10s.

5. 8379 lies × 7 tablespoons/lie = 58,653 tablespoons of habanero sauce. This could be some people's definition of hell.

6. $\frac{25}{65} = \frac{5}{13}$ I divided top and bottom by 5

7.
$$\begin{array}{r} 2{,}000{,}000{,}000 \\ +\,400{,}000{,}000{,}000 \\ \hline 402{,}000{,}000{,}000 \end{array}$$

The Bridge
answers

8. \$8/hour × 24 hours/day = \$192/day
\$192/day × 7 days/week = \$1,344/week.

9. 29,028' + 26,503' = 55,531'

10. 64' 9"

$$\begin{array}{r} 64\text{ R }9 \\ 12\overline{)777} \\ -\underline{72} \\ 57 \\ -\underline{48} \\ 9 \end{array}$$

from p. 55 — *fourth try*

1. 13 pounds × 16 ounces/pound = 208 ounces.

2. one-eighth is larger. (See the diagram.)

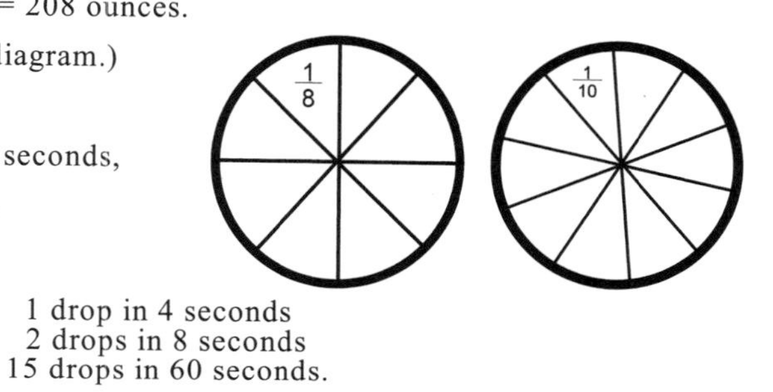

3. If he loses one drop every four seconds, he will lose 15 drops each minute. How do I know this?

One way is to start counting: 1 drop in 4 seconds
2 drops in 8 seconds
until I get to . . . 15 drops in 60 seconds.

The easier way is to realize that there are 15 four-second periods in 60 seconds. $4\overline{)60}$ = 15

Then: $\frac{15\text{ drops}}{1\text{ }\cancel{\text{minute}}} \times \frac{4\text{ }\cancel{\text{minutes}}}{1}$ = 60 drops in four minutes.

Or better yet, $\frac{1\text{ drop}}{4\text{ }\cancel{\text{secs}}} \times \frac{60\text{ }\cancel{\text{secs}}}{1\text{ }\cancel{\text{min}}} \times \frac{4\text{ }\cancel{\text{minutes}}}{1}$ = 60 drops in four minutes.

4. $\frac{120}{360} = \frac{1}{3}$ You could take many steps to do this, or just divide top and bottom by 120 to do it in one step.

5. 3 → 6 → 12 → 24 → 48 → 96 → 192 → 384 → 768 → 1536 → 3072

6. $\frac{1}{12} + \frac{5}{12} = \frac{6}{12}$ which reduces to $\frac{1}{2}$

7. $\$1,000,000,000 \le x$

8. Half of the size of the Sahara is ½ × 3,500,000 = 1,750,000.
The Great Australian is only 1,480,000, which is less than half of the size of the Sahara.

9. 7 furlongs × 220 yards/furlong × 3 feet/yard = 4,620 feet

10.
$$\begin{array}{r} 55,000,000 \\ +\ \underline{26,000,000,000} \\ 26,055,000,000 \end{array}$$

The Bridge
answers

from p. 56 — *fifth try*

1. 8 minutes × 60 seconds/minute = 480 seconds
2. 146 + 677 = 823
3. 10,000,000,000. You need ten 10s multiplied together to get ten billion.
4. $\frac{3}{14} + \frac{4}{14} = \frac{7}{14}$ which reduces to $\frac{1}{2}$
5. 125 pounds × 16 ounces/pound = 2000 ounces.
6. 78" = 6' 6" $\quad 12\overline{)78}$ = 6 R 6; $78 - 72 = 6$
7. (See diagram →)

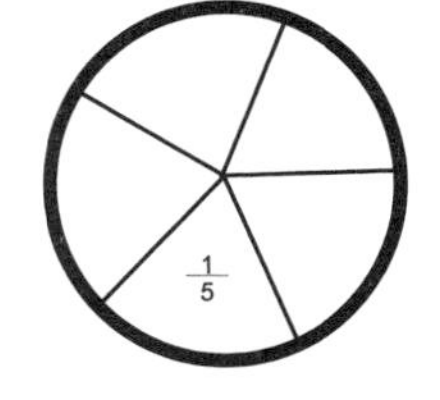

8. Zero times any number is zero.
9. $\frac{30}{60} = \frac{1}{2}$
10. \$493/year × 100 years/century = \$49,300.

from p. 74 — *first try*

1. $\frac{16}{16} - \frac{3}{16} = \frac{13}{16}$
2. $\frac{5}{20}$ which reduces to $\frac{1}{4}$
3. $\frac{1}{3} = \frac{7}{21}$ and $\frac{2}{7} = \frac{6}{21}$ so $\frac{1}{3} > \frac{2}{7}$ $\quad \frac{2}{7}$ is smaller
4. 39 = XXXIX
5. XLIX $\qquad$ XL is the 40; IX is the 9
6. $\frac{2}{3} + \frac{1}{4} = \frac{8}{12} + \frac{3}{12} = \frac{11}{12}$
7. The whole numbers are {0, 1, 2, 3, 4. . .}. The only whole number less than 1 is 0.
8. $\frac{2}{5} = \frac{6}{15}$ and $\frac{1}{3} = \frac{5}{15}$ so $\frac{2}{5} < \frac{1}{3}$ is false.
9. $1000000 - 198 = 999802$ $\qquad$ \$999,802
10. One-fourth of 360° = ¼ × 360 = $4\overline{)360}$ = 90 = 90°

from p. 75 — *second try*

1. $\frac{3}{9}$ is not reduced. Joe should have written $\frac{1}{3}$ as his final answer.
2. LXIV ÷ XVI becomes 64 ÷ 16 which equals 4. 4 = IV
3. Yes, since ≥ means either greater than OR equal to, and $7\frac{1}{3}$ does equal $7\frac{1}{3}$.

The Bridge
answers

4. $\frac{8}{8} - \frac{3}{8} = \frac{5}{8}$

5. The first double gives you 14. Then 28, 56, 112, 224, 448, 896, 1792

6. $\frac{1}{3} + \frac{2}{5} = \frac{5}{15} + \frac{6}{15} = \frac{11}{15}$

7. $16 \div 32 = \frac{16}{32}$ which reduces to $\frac{1}{2}$

8. The radius is half of the diameter. Half of 22" is 11".

9. $12\overline{)77}$ = 6 R 5, $77 - 72 = 5$ 77" = 6' 5"

10. $\frac{2}{5} = \frac{4}{10}$ and $\frac{1}{2} = \frac{5}{10}$ Therefore, $\frac{2}{5} < \frac{1}{2}$ $\frac{1}{2}$ is larger

from p. 76 — *third try*

1. $\frac{3}{5} + \frac{3}{8} = \frac{24}{40} + \frac{15}{40} = \frac{39}{40}$

2. LXXIV + XVI = 74 + 16 = 90 = XC

3. $\frac{16}{16} - \frac{7}{16} = \frac{9}{16}$

4. $\frac{1}{3} = \frac{10}{30}$ and $\frac{3}{10} = \frac{9}{30}$ Therefore, $\frac{1}{3} > \frac{3}{10}$ and $\frac{3}{10}$ is smaller.

5. One-half of 360° = ½ × 360° = 180°

6. $\frac{204}{276}$ The numerator and denominator are both even so we can divide both by 2.

$\frac{204}{276} = \frac{102}{138}$ They are still both even. Divide by 2 again. $\frac{102}{138} = \frac{51}{69}$

What divides into both 51 and 69? Two won't work. Let's try three. $\frac{51 \div 3}{69 \div 3} = \frac{17}{23}$

7. $\frac{11}{12} - \frac{1}{12} = \frac{10}{12} = \frac{5}{6}$

8. Fifty-five million, four hundred forty-four thousand, nine hundred ninety-nine

9. The first double gives you 18. Then 36, 72, 144, 288, 576, 1152, 2304

10. Five sandwiches divided equally among 15 ducks gives each duck $\frac{5}{15}$ of a sandwich. This reduces to $\frac{1}{3}$

from p. 77 — *fourth try*

1. 7 is a larger numeral than 9. (But 9 is a larger number than 7.)

2. In order to change $\frac{67}{90}$ into a fraction with a denominator of 720, we have to multiply top and bottom by 8. $\frac{67 \times 8}{90 \times 8} = \frac{536}{720}$ The numerator is 536.

3. $\frac{6}{8}$ is not reduced. He should have written $\frac{3}{4}$

4. $\frac{1}{10} + \frac{1}{40} = \frac{4}{40} + \frac{1}{40} = \frac{5}{40}$ which reduces to $\frac{1}{8}$

5. A diameter is twice the radius. 2 × 88' = 176'

6. $\frac{5}{6} = \frac{35}{42}$ and $\frac{6}{7} = \frac{36}{42}$ Therefore, $\frac{5}{6} < \frac{6}{7}$

7. $12\overline{)88}$ = 7 R 4; 88 − 84 = 4   88" = 7' 4"

8. $\frac{8}{8} - \frac{5}{8} = \frac{3}{8}$

9. CXXVI ÷ XIV = 126 ÷ 14 = 9 = IX

10. One-eighth of 360° = $\frac{1}{8} \times 360° = 8\overline{)360}$ = 45 = 45°

from p. 78 — *fifth try*

1. 1000000000 − 3489 = 999996511 $999,996,511

2. $\frac{10}{10} - \frac{7}{10} = \frac{3}{10}$

3. Three cups divided among four pancakes = 3 ÷ 4 = $\frac{3}{4}$ cup

4. In order to change $\frac{87}{90}$ into a fraction with a denominator of 630, we have to multiply top and bottom by 7. $\frac{87 \times 7}{90 \times 7} = \frac{609}{630}$ The numerator is 609.

5. $\frac{2}{3} = \frac{8}{12}$ and therefore, $\frac{2}{3} > \frac{7}{12}$

6. IX + XCIX = 9 + 99 = 108 = CVIII

7. $\frac{1}{9} + \frac{1}{99} = \frac{11}{99} + \frac{1}{99} = \frac{12}{99}$ $\frac{12 \div 3}{99 \div 3} = \frac{4}{33}$

8. 20000 − 589 = 19411   All she needs is $19,411 more.

9. The whole numbers are {0, 1, 2, 3, 4, 5. . .}. Zero and one are the two whole numbers less than two.

10. $\frac{11}{12} - \frac{1}{12} = \frac{10}{12}$ $\frac{10 \div 2}{12 \div 2} = \frac{5}{6}$

from p. 90 — *first try*

1. $5280\overline{)1000000}$ = 189 R 2080
   1000000 − 5280 = 47200
   47200 − 42240 = 49600
   49600 − 47520 = 2080

   189 miles 2080 feet

2. The LCM of 6 and 8 is 24.

The Bridge
answers

3. *First* is an ordinal number, not a cardinal number.

4. $\frac{4}{4}$ and $\frac{9}{4}$ are examples of improper fractions. In each case, the denominator is ≤ the numerator.

5. One-seventh of \$5523 = $\begin{array}{r} 789 \\ 7\overline{)\,5523} \end{array}$ He has \$789.

6. There are nine lines of symmetry. I'll draw one of them →

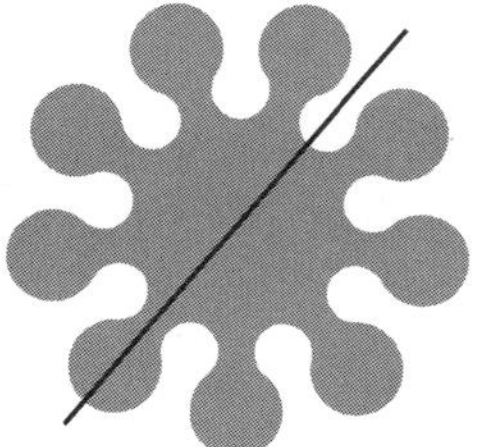

7. Any line through the center of a circle is a line of symmetry. There is an unlimited number of lines of symmetry for a circle: an infinite number.

8. The number of 8s you have to add up in order to get 6,232 is the same as $\frac{6232}{8}$ which is the same as $\begin{array}{r} 779 \\ 8\overline{)\,6232} \end{array}$

9. Seven billion, nine hundred forty-four thousand, three hundred thirty-three.

10. Any number which is less than or equal to 44 will make "$y \leq 44$" and "$y < 62$" both true.

from p. 91 — *second try*

1. $\begin{array}{r} 15 \text{ R } 2 \\ 4\overline{)\,62} \\ -4 \\ \hline 22 \\ -20 \\ \hline 2 \end{array}$ 62 quarts = $15\frac{2}{4} = 15\frac{1}{2}$ gallons

2. Answers will vary. For example, $5\frac{1}{2}$ or $397\frac{2}{3}$ or $1\frac{204}{276}$

A mixed number is an integer followed by a fraction.

3. $\frac{12}{6}$ which equals 2.

4. Answers will vary. Many readers will draw a seven-pointed blob similar to the nine-pointed blob in question 6 of "first try," but there are millions of other correct drawings. Here's another possibility.

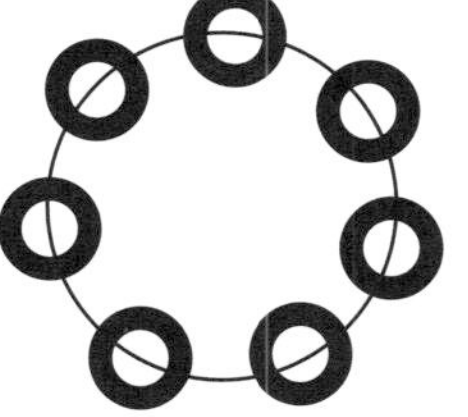

5. $\begin{array}{r} 125 \\ 8\overline{)\,1000} \end{array}$ 1000 fluid ounces = 125 cups.

6. A sector has one line of symmetry. (See diagram →)

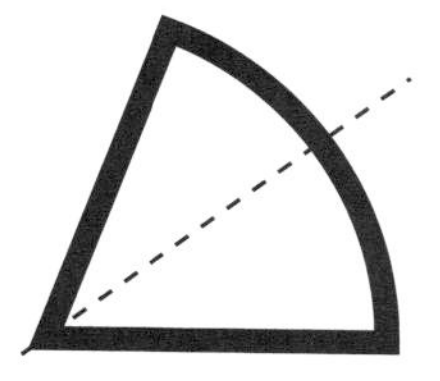

7. 12 × \$17 = \$204.

8. 59 = LIX 102 = CII

9. The LCM of 8 and 10 is 40.

10. There are no lines of symmetry.

The Bridge
answers

from p. 92 — *third try*

1. $\frac{7}{8} + \frac{7}{9}$ has a LCD of 72.

$\frac{7}{8} = \frac{63}{72}$ (Multiplying top and bottom by 9.)

$\frac{7}{9} = \frac{56}{72}$ (Multiplying top and bottom by 8.)

$\frac{63}{72} + \frac{56}{72} = \frac{119}{72} = 1\,\frac{47}{72}$

2. One-eighth of 632 = $\begin{array}{r} 79 \\ 8\overline{)632} \end{array}$ His bicycle box weighs 79 pounds.

3. DCXXIII ÷ VII = 623 ÷ 7 = 89 = LXXXIX

4. The smallest number that both a million and a billion evenly divide into is a billion.

5. $\frac{8}{9}$

6. Such a fraction is called an improper fraction.

7. 7 miles × 5280 feet/mile = 36,960 feet

8. Eight million, two hundred forty-seven thousand, nine hundred twenty-one.

9. *First* is the smallest ordinal number.

10. There are no lines of symmetry for that trapezoid.

from p. 93 — *fourth try*

1. One line of symmetry.

2. The smallest number that 20 and 20,000 both evenly divide into is 20,000.

3. There is no number that is both ≥ 100 and < 50.

4. 7, 8, and 9 all make $7 \leq x \leq 9$ true.

5. The number of 7s you would have to add up to get 62,629 = $\frac{62629}{7}$ = $\begin{array}{r} 8947 \\ 7\overline{)62629} \end{array}$

6. The least common denominator of $\frac{1}{6}$ and $\frac{1}{8}$ is 24.

7. $\begin{array}{r} 9 \text{ R } 3 \\ 12\overline{)111} \end{array}$ 111" = 9' 3"

8. 111" = $9\,\frac{3}{12} = 9\,\frac{1}{4}$ feet

9. $\frac{5}{9}$

10. $\begin{array}{r} 16666 \text{ R } 40 \\ 60\overline{)1000000} \end{array}$ 1,000,000 minutes = 16,666 hours, 40 minutes

The Bridge
answers

from p. 94 — *fifth try*

1. One-sixth of 534 = $6\overline{)534}$ = 89 Each bag weighs 89 ounces.
2. Since a radius is half the length of the diameter, it would be 25 furlongs.
3. $\frac{18}{8} = 2\frac{2}{8} = 2\frac{1}{4}$
4. VII × LXXVII = 7 × 77 = 539 = DXXXIX
5. $4\overline{)3157}$ = 789 R 1 789 gallons 1 quart
6. It had one line of symmetry. (See diagram →)

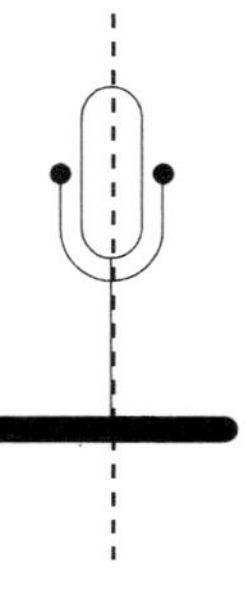

7. 33,000,000,926
8. Here are the General Rules:
   (1) Reduce fractions in your answers as much as possible.
   (2) Fractions like $\frac{0}{4}$ are equal to 0.
   (3) Fractions like $\frac{4}{4}$ are equal to 1.
   (4) Division by zero doesn't make sense.

By (2), $\frac{0}{2} = 0$, which is perfectly fine.

9. The smallest number that 3, 4, 5, and 6 evenly divide into is 60.
10. The least common denominator of $\frac{1}{4}$, $\frac{1}{5}$ and $\frac{1}{8}$ is 40.

from p. 108 — *first try*

1. 17 times/minute × 60 minutes/hour = 1020 times/hour
2. $15\frac{1}{7}$ oz./day × 7 days/week = $\frac{106}{7} \times \frac{7}{1} = \frac{742}{7} = 7\overline{)742}$ = 106 = 106 oz./week
3. There are many possible numbers that make $5 < y < 6$ true. For example, $5\frac{1}{2}$ or $5\frac{1}{8}$
4. The least common multiple of 7, 14 and 28 is 28.
5. $17\overline{)1515}$ = 89 R 2 $89\frac{2}{17}$
6. The integers = { . . . –3, –2, –1, 0, 1, 2 . . . } and the whole numbers = {0, 1, 2 . . . } so numbers like –2 are integers that are not whole numbers.
7.
$$\begin{array}{rrcl} & 148\frac{1}{2} & = & 148\frac{4}{8} \\ + & 1\frac{7}{8} & = & 1\frac{7}{8} \\ \hline & & & 149\frac{11}{8} = 149 + \frac{8}{8} + \frac{3}{8} = 150\frac{3}{8} \text{ lbs.} \end{array}$$

8. $\frac{5}{8} + \frac{5}{8} = \frac{10}{8} = \frac{8}{8} + \frac{2}{8} = 1\frac{2}{8} = 1\frac{1}{4}$ lbs.

The Bridge
answers

9. If the radius is equal $8\frac{1}{4}$ then the diameter equals $16\frac{1}{2}$ (which is $8\frac{1}{4} + 8\frac{1}{4}$)

$3\frac{1}{7} \times 16\frac{1}{2} = \frac{22}{7} \times \frac{33}{2} = \frac{726}{14} = 14\overline{)726}$ (51 R12) $= 51\frac{12}{14} = 51\frac{6}{7}$ inches

10. $1{,}000 \times 1{,}000{,}000 = 1{,}000{,}000{,}000$.

from p. 109 — *second try*

1. $73\frac{1}{3}$ = *3 times 73 . . . plus 1* $= \frac{220}{3}$

2.
$$\begin{array}{rl} 8\frac{7}{8} = & 8\frac{21}{24} \\ -\ 4\frac{2}{3} = & 4\frac{16}{24} \\ \hline & 4\frac{5}{24} \text{ oz.} \end{array}$$

3. $\frac{42}{54} = \frac{42 \div 6}{54 \div 6} = \frac{7}{9}$

4. $\frac{3}{4} + \frac{1}{8} + \frac{1}{2} = \frac{6}{8} + \frac{1}{8} + \frac{4}{8} = \frac{11}{8} = 1\frac{3}{8}$ lbs.

5. XXI, XXII, XXIII, XXIV, XXV, XXVI, XXVII, XXVIII, XXIX, XXX.

6. The least common multiple of 6 and 9 is 18.

7. Answers will vary. For example:

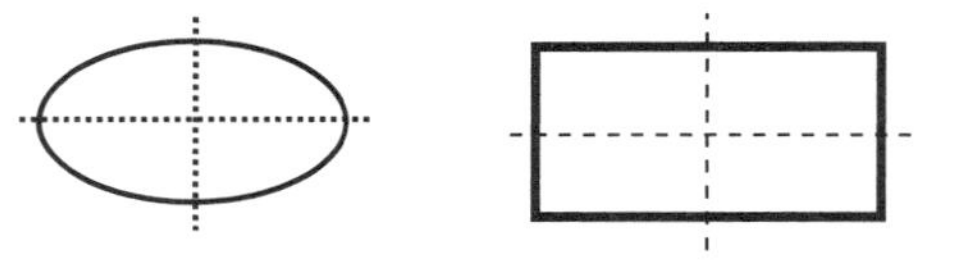

8. Fifty-seven billion, nine hundred eighty-three million.

9. I want more money in my pocket than the cost of the oboe. I want $x > 143$.

10. $\frac{992}{34} = 34\overline{)992}$ (29 R 6) $= 29\frac{6}{34} = 29\frac{3}{17}$

from p. 110 — *third try*

1. $\frac{1}{4} + \frac{1}{8} + 2\frac{1}{2} = \frac{2}{8} + \frac{1}{8} + 2\frac{4}{8} = 2\frac{7}{8}$ oz.

2. $3\frac{1}{7} \times 11\frac{1}{10} = \frac{22}{7} \times \frac{111}{10} = \frac{2442}{70} = 70\overline{)2442}$ (34 R 62) $= 34\frac{62}{70} = 34\frac{31}{35}$ feet

3.
$$\begin{array}{rl} 55\frac{1}{8} = & 55\frac{1}{8} \\ +\ 27\frac{3}{4} = & 27\frac{6}{8} \\ \hline & 82\frac{7}{8} \end{array}$$

The Bridge
answers

4. *Third* is an ordinal number.

5. 4 is a cardinal number.

6. $\frac{3}{4} \times \frac{7}{6} = \frac{21}{24} = \frac{7}{8}$

7. 5 times/minute × 60 minutes/hour × 6 hours = 1800 times in six hours

8. $$\begin{array}{rcl} & \frac{6}{7} = & \frac{18}{21} \\ - & \frac{2}{3} = & \frac{14}{21} \\ \hline & & \frac{4}{21} \end{array}$$

9. CXXXVI ÷ XVII = 136 ÷ 17 = 8 = VIII

10. $\frac{276}{38} = 38\overline{)276}^{\;7\text{ R }10} = 7\frac{10}{38} = 7\frac{5}{19}$

from p. 111 — *fourth try*

1. $\frac{287}{71} = 71\overline{)287}^{\;4\text{ R }3} = 4\frac{3}{71}$

2. She could do them in either order and still get the same result. It is commutative.

3. $$\begin{array}{rcl} & 6\frac{2}{3} = & 6\frac{4}{6} \\ - & 5\frac{1}{2} = & 5\frac{3}{6} \\ \hline \end{array}$$
$1\frac{1}{6}$ oz. is the amount he left in the bottle

4. 278" = $12\overline{)278}^{\;23\text{ R }2}$ = 23' 2"

5. $1000\frac{2}{3}$ = *three times 1000 . . . plus 2 =* $\frac{3002}{3}$

6. The LCM of 20 and 30 is 60.

7. Answers will vary. For example:

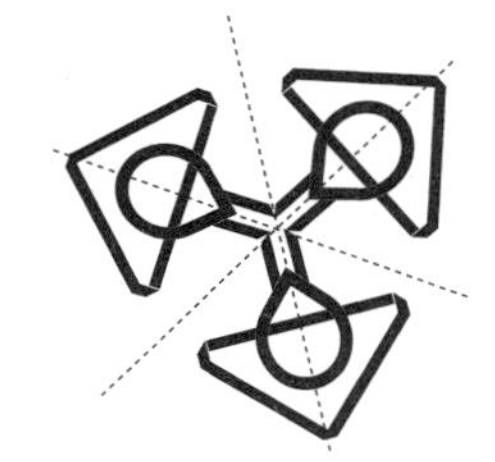

8. $\frac{1}{2}$ of 360° = 180°

9. 8 feet/second × 60 seconds = 480 feet

10. Seven and one-fifth.

from p. 112 — *fifth try*

1. The smallest number that 2, 3, 4, and 8 evenly divide into is 24.

2. $3\frac{1}{7} \times 2\frac{3}{4} = \frac{22}{7} \times \frac{11}{4} = \frac{242}{28} = 28\overline{)242}^{\;8\text{ R }18} = 8\frac{18}{28} = 8\frac{9}{14}$ miles

The Bridge
answers

3. 17 oz./day × 7 days = 119 oz.
4. Answers may vary. z could be $10\frac{1}{2}$ or $10\frac{1}{8}$ or $10\frac{98}{99}$
5. $\frac{181}{44} = 44\overline{)181}$ = 4 R 5 $= 4\frac{5}{44}$
6. $6\frac{1}{6} = 6\frac{1}{6}$

   $+\ 7\frac{1}{3} = 7\frac{2}{6}$

   $13\frac{3}{6} = 13\frac{1}{2}$
7. $\frac{4}{5} = \frac{28}{35}$

   $-\ \frac{3}{7} = \frac{15}{35}$

   $\frac{13}{35}$
8. $1 - \frac{3}{5} = \frac{5}{5} - \frac{3}{5} = \frac{2}{5}$
9. one-quarter of 16 oz. $= \frac{1}{4} \times 16 = \frac{1}{4} \times \frac{16}{1} = \frac{16}{4} = 4$ oz.

10. There are two possible answers. You could say that since she drank one-quarter of the milkshake, she didn't drink three-quarters of it. Or you could say that since she drank four ounces of the sixteen-ounce milkshake, she didn't drink twelve ounces.

from p. 126 — *first try*

1. $\frac{5}{66}$
2. $\frac{7 \text{ jobs}}{\text{minute}} \times \frac{60 \text{ minutes}}{1 \text{ hour}} \times 8 \text{ hours} = 3{,}360 \text{ jobs}$
3. One-third of \$534 = $3\overline{)534}$ = 178 = \$178
4. $\frac{1}{10} \times 5\frac{1}{5} = \frac{1}{10} \times \frac{26}{5} = \frac{13}{25}$
5. $\frac{3}{7} = \frac{27}{63}$ $\frac{5}{9} = \frac{35}{63}$ So $\frac{3}{7} < \frac{5}{9}$
6. $3\frac{1}{3} \times 3\frac{1}{3} = \frac{10}{3} \times \frac{10}{3} = \frac{100}{9} = 11\frac{1}{9}$
7. $19\frac{1}{3} \times 40 = 773\frac{1}{3}$
8. 9 × 9 = 81 So 9 is the square root of 81.
9. $5\frac{1}{5} - 3\frac{1}{3} = 5\frac{3}{15} - 3\frac{5}{15} = 4\frac{15}{15} + \frac{3}{15} - 3\frac{5}{15} = 1\frac{13}{15}$
10. $8\frac{1}{2} + \frac{1}{3} = 8\frac{3}{6} + \frac{2}{6} = 8\frac{5}{6}$ oz.

The Bridge
answers

from p. 127 — *second try*

1. $\frac{7}{10} = \frac{28}{40}$ $\frac{5}{8} = \frac{25}{40}$ So $\frac{5}{8} < \frac{7}{10}$

2. Start with $\frac{1}{90}$ After the first double we have $\frac{2}{90}$ which is $\frac{1}{45}$

After the second double $\frac{2}{45}$ Third double $\frac{4}{45}$ Fourth double $\frac{8}{45}$ Fifth $\frac{16}{45}$

Sixth $\frac{32}{45}$ Seventh $\frac{64}{45}$ which is $1\frac{19}{45}$ It took seven doublings.

3. $35 - 3\frac{1}{8} = 34\frac{8}{8} - 3\frac{1}{8} = 31\frac{7}{8}$

4. Nine tens since there are nine zeros in 1,000,000,000.

5. $\frac{\overset{1}{\cancel{6}}}{\underset{5}{\cancel{25}}} \times \frac{\overset{3}{\cancel{15}}}{\underset{4}{\cancel{24}}} = \frac{3}{20}$

6. Subtract seven and then divide by four.

7. $\frac{\text{20 Darlene's cookies}}{1} \times \frac{\text{2 Joe's cookies}}{\text{9 Darlene's cookies}} = \frac{40}{9} = 4\frac{4}{9}$ of Joe's cookies

8. 667 = DCLXVII

9. 1,000,000 – 67 = 999,933

10. $\frac{\text{6 Joe's cookies}}{\text{8 people}} = \frac{6}{8} = \frac{3}{4}$ of a cookie per person

from p. 128 — *third try*

1. $\frac{5}{6}$

2. You would multiply by 1,000.

3. $2\frac{3}{4} \times 8\frac{1}{8} = \frac{11}{4} \times \frac{65}{8} = \frac{715}{32} = 32\overline{)715}$ (22 R11) $= 22\frac{11}{32}$ square inches

4. $7\frac{1}{7} - 2\frac{5}{7} = 6\frac{7}{7} + \frac{1}{7} - 2\frac{5}{7} = 4\frac{3}{7}$

5. The whole numbers are {0, 1, 2, 3, . . . }, and if we add any two of them we get another whole number. And since every whole number is an integer = {. . . –2, –1, 0, 1, 2, 3, . . . }, it is true that the sum of any two whole numbers is an integer.

6. 819

7. $\frac{12}{60} = \frac{1}{5}$

8. $\frac{1}{7} \times 8\frac{2}{5} = \frac{1}{7} \times \frac{42}{5} = \frac{42}{35} = 1\frac{7}{35} = 1\frac{1}{5}$ or $\frac{1}{\underset{1}{\cancel{7}}} \times \frac{\overset{6}{\cancel{42}}}{5} = \frac{6}{5} = 1\frac{1}{5}$

9. Multiply by $\frac{3}{7}$ . . . or divide by $\frac{7}{3}$

10. $\frac{\text{Darlene's 14 pieces}}{1} \times \frac{\text{Joe's 19 pieces}}{\text{Darlene's 2 pieces}} = 133$ Joe's pieces

The Bridge
answers

from p. 129 — *fourth try*

1. $\frac{3}{4} = \frac{9}{12}$ $\frac{2}{3} = \frac{8}{12}$ Therefore, $\frac{3}{4} > \frac{2}{3}$

2. 2,000,000,000 – 1,000,000 = 1,999,000,000

3. $45\frac{1}{5} - 20 = 25\frac{1}{5}$

4. $\frac{2}{3}$

5. $\frac{\$500}{1 \text{ month}} \times \frac{12 \text{ months}}{1 \text{ year}} \times 10 \text{ years} = \$60{,}000.$

6. $\frac{3}{7} = \frac{15}{35}$

7. 539 = DXXXIX

8. Canceling makes this problem easier. $\frac{3}{5}$

9. $\frac{87 \text{ tables}}{1} \times \frac{45 \text{ seconds}}{1 \text{ table}} = 3915 \text{ seconds.}$ $60\overline{)3915}$ = 65 R 15

= 65 minutes 15 seconds

10. One-seventh of 476 = $7\overline{)476}$ = 68 = 68" = 5' 8"

from p. 130 — *fifth try*

1. $\frac{1}{12}$

2. CVI + XC = 106 + 90 = 196 = CXCVI

3. $4\frac{1}{4}$

4. $4\frac{1}{4} \times 4\frac{1}{4} = \frac{17}{4} \times \frac{17}{4} = \frac{289}{16}$ $16\overline{)289}$ = 18 R 1 $18\frac{1}{16}$

5. $\frac{77 \text{ feet}}{1 \text{ minute}} \times 5 \text{ minutes} = 385 \text{ feet}$

6. $8\frac{1}{3}$

7. $\frac{7}{8} = \frac{56}{64}$ so $\frac{7}{8} < \frac{57}{64}$

8. A radius is half of a diameter. $\frac{1}{2}$ of 17 = $2\overline{)17}$ = 8 R 1 = $8\frac{1}{2}$ inches

9. There is no cardinal number x such that $17 < x < 18$. The cardinal numbers are used for counting things like forks or mice. You can't have fractional amounts.

10. Ten inches times one side of the rectangle will equal the area which is 360 square inches. What do you multiply 10 by in order to get 360? You multiply by 36. So the windshield is 36" wide.

from p. 142 — *first try*

1. $8\frac{1}{2} \times 11 = 93\frac{1}{2}$ square inches

2. $30\frac{1}{28}$

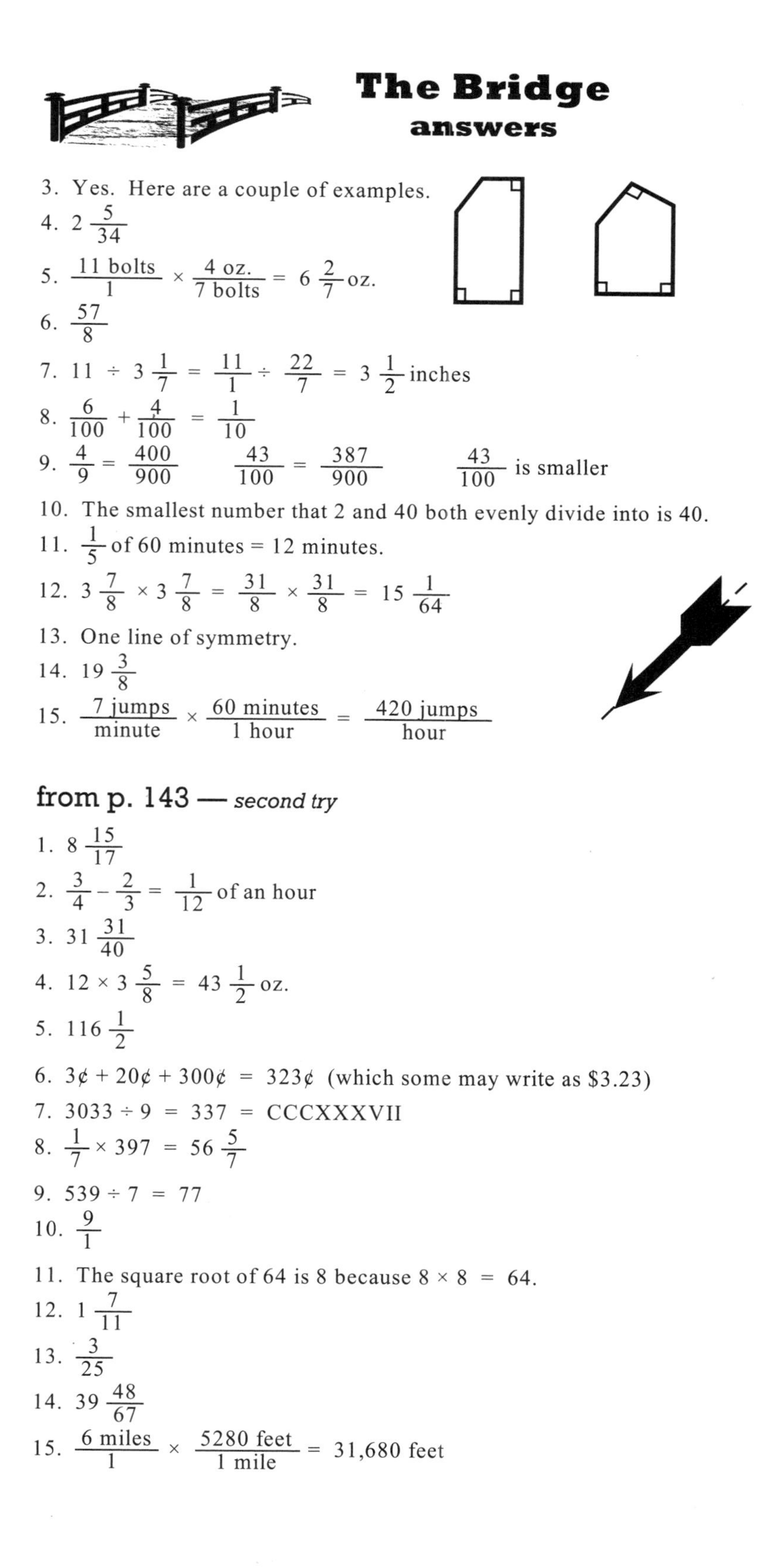

The Bridge
answers

3. Yes. Here are a couple of examples.
4. $2\frac{5}{34}$
5. $\frac{11 \text{ bolts}}{1} \times \frac{4 \text{ oz.}}{7 \text{ bolts}} = 6\frac{2}{7}$ oz.
6. $\frac{57}{8}$
7. $11 \div 3\frac{1}{7} = \frac{11}{1} \div \frac{22}{7} = 3\frac{1}{2}$ inches
8. $\frac{6}{100} + \frac{4}{100} = \frac{1}{10}$
9. $\frac{4}{9} = \frac{400}{900}$ $\quad \frac{43}{100} = \frac{387}{900}$ $\quad \frac{43}{100}$ is smaller
10. The smallest number that 2 and 40 both evenly divide into is 40.
11. $\frac{1}{5}$ of 60 minutes = 12 minutes.
12. $3\frac{7}{8} \times 3\frac{7}{8} = \frac{31}{8} \times \frac{31}{8} = 15\frac{1}{64}$
13. One line of symmetry.
14. $19\frac{3}{8}$
15. $\frac{7 \text{ jumps}}{\text{minute}} \times \frac{60 \text{ minutes}}{1 \text{ hour}} = \frac{420 \text{ jumps}}{\text{hour}}$

from p. 143 — *second try*

1. $8\frac{15}{17}$
2. $\frac{3}{4} - \frac{2}{3} = \frac{1}{12}$ of an hour
3. $31\frac{31}{40}$
4. $12 \times 3\frac{5}{8} = 43\frac{1}{2}$ oz.
5. $116\frac{1}{2}$
6. 3¢ + 20¢ + 300¢ = 323¢ (which some may write as $3.23)
7. $3033 \div 9 = 337$ = CCCXXXVII
8. $\frac{1}{7} \times 397 = 56\frac{5}{7}$
9. $539 \div 7 = 77$
10. $\frac{9}{1}$
11. The square root of 64 is 8 because $8 \times 8 = 64$.
12. $1\frac{7}{11}$
13. $\frac{3}{25}$
14. $39\frac{48}{67}$
15. $\frac{6 \text{ miles}}{1} \times \frac{5280 \text{ feet}}{1 \text{ mile}} = 31{,}680$ feet

The Bridge
answers

from p. 144 — *third try*

1. $\frac{1}{21}$
2. $5\frac{1}{4}$ quarts $= \frac{21 \text{ quarts}}{4} \times \frac{1 \text{ gallon}}{4 \text{ quarts}} = 1\frac{5}{16}$ gallons
3. $14\frac{1}{3}$
4. There are no lines of symmetry.
5. The smallest number that 3 and 40 evenly divide into is 120.
6. $\frac{1}{6} \times \frac{31}{6} = \frac{31}{36}$
7. Answers will vary. For example, 45 ÷ 89 doesn't equal 89 ÷ 45.
8. 25 square miles
9. $999{,}999{,}999\frac{1}{8}$
10. $52\frac{1}{7} \div 10 = 5\frac{3}{14}$ pounds
11. $41\frac{43}{49}$
12. $\frac{3}{7} + \frac{43}{100} = \frac{300}{700} + \frac{301}{700} = \frac{601}{700}$
13. Looking at the previous problem makes this problem easy. $\frac{43}{100}$ is larger than $\frac{3}{7}$
14. Nine diagonals
15. $18,700 ÷ 5 = $3,740.

from p. 145 — *fourth try*

1. $5\frac{1}{25}$
2. $\frac{1}{12} \times 60$ minutes $=$ 5 minutes
3. $13\frac{13}{18}$
4. No. No. Yes.
5. $\frac{1}{12}$
6. $12\frac{9}{64}$ square feet
7. 99 × 9 = 891 = DCCCXCI
8. 2
9. $5\frac{1}{4}$ gallons $= \frac{21 \text{ gallons}}{4} \times \frac{4 \text{ quarts}}{1 \text{ gallon}} =$ 21 quarts
10. C) should equal at least 14 since both the numbers added are greater than 7.
    D) should be less than 24 since $6 \times 4 = 24$ and $5\frac{1}{5} < 6$ and $3\frac{3}{5} < 4$.
11. The inverse to "take the square root of " is "square the number."
12. $3\frac{7}{12}$

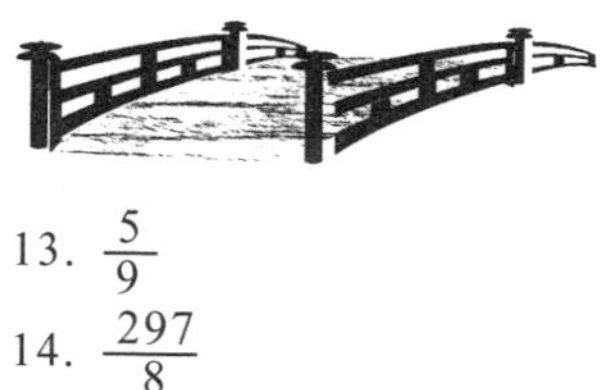

The Bridge
answers

13. $\frac{5}{9}$
14. $\frac{297}{8}$
15. $1 - \frac{13}{16} = \frac{3}{16}$

from p. 146 — *fifth try*

1. 10 is the square root of 100 since $10 \times 10 = 100$.
2. $35 \frac{6}{19}$
3. $\frac{1}{44}$ furlong $= \frac{1 \text{ furlong}}{44} \times \frac{220 \text{ yards}}{1 \text{ furlong}} = 5$ yards
4. $44 \frac{1}{8}$
5. Five diagonals
6. $1 \frac{7}{8}$
7. $\frac{5}{11}$
8. $\frac{1}{3}$
9. $\frac{3}{5}$ since $\frac{3}{5} = \frac{60}{100}$
10. $a + b = b + a$ is the commutative law of addition.
11. The square of 17 means 17×17 which is 289.
12. Any number (except zero) divided by itself equals one.
13. Any number multiplied by zero equals zero.
14. It has four lines of symmetry.
15. $\frac{1 \text{ week}}{8} \times \frac{7 \text{ days}}{1 \text{ week}} \times \frac{24 \text{ hours}}{1 \text{ day}} = 21$ hours

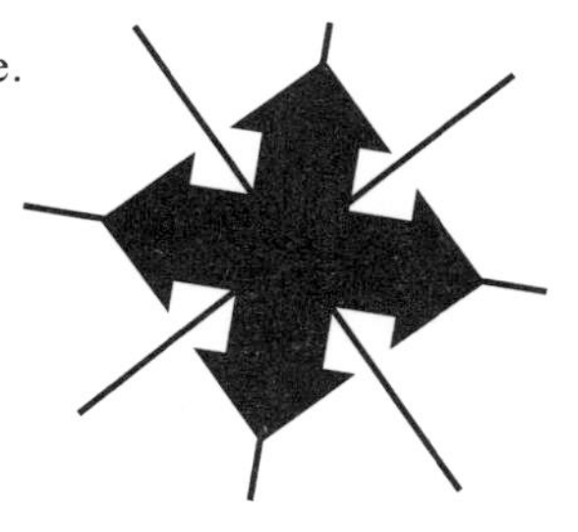

from page 17

1. *The symbol $<$ stands for "is less than."*
2. $88 < 92$ is true.
3. $100 < 12$ is false.
4. $5 < 5\frac{1}{2}$ is true.
5. Your answer might be different from mine. I wrote $14 < 397972$. You may have written 15 or 16 or 17 or 14½ or 14.001. Any of these is correct.
6. Your answer might be different from mine. I wrote $2\frac{1}{2} < 3$. You may have written 0 or 1 or 2.
7. Most people put one number on top of the other (arrange them vertically) before they add them.

$$\begin{array}{r} 389 \\ +\ 772 \\ \hline 1161 \end{array}$$

8. $>$ means the opposite of $<$.

$<$ means "is less than."
$>$ means "is greater than."

All of these are true:

$8 > 4$
$1089723949237 > 2$
$2 < 1089723949237$
$9 > 0$
$5\frac{1}{2} > 5$

from page 20

1. Since there are 60 minutes in an hour, he could list sixty times as many reasons.

$$\begin{array}{r} 60 \\ \times\ 60 \\ \hline 3600 \end{array}$$

which some people write as→

$$\begin{array}{r} 60 \\ \times\ 60 \\ \hline 00 \\ 360 \\ \hline 3600 \end{array}$$

2. We need to take the 3600 for each hour and multiply by 24 hours. He could list 86,400 reasons per day.

$$\begin{array}{r} 3600 \\ \times\ \ 24 \\ \hline 14400 \\ 7200 \\ \hline 86400 \end{array}$$

3. We need to take the 86,400 reasons per day and multiply by 365. So, there are 31,536,000 seconds in a year. Written out, that would be thirty-one million, five hundred thirty-six thousand. It's important to learn how to write out numbers in words, because every time you write a check, you need to do that.

$$\begin{array}{r} 86400 \\ \times\ \ 365 \\ \hline 432000 \\ 518400 \\ 259200 \\ \hline 31536000 \end{array}$$

Fred wishes that he could write a check like this:

4. One last multiplication: 31 × 31,536,000.

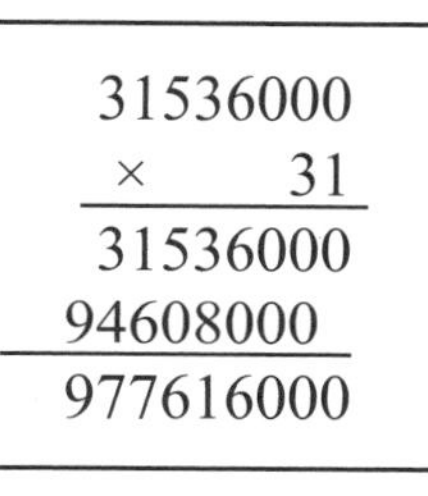

```
  31536000
 ×      31
  31536000
 94608000
 977616000
```

We notice that 977,616,000 is just a little less than 1,000,000,000. Therefore, it would take more than 31 years for Fred to list a billion reasons why he should own a bike. In 32 years, he could do it. When I multiply the number of seconds in a year (31,536,000) by 32, I get 1,009,152,000 (one billion, nine million, one hundred fifty-two thousand).

from page 23

1. 1,000,000,000 is a cardinal number. A company that sells hamburgers once said that it had just served its billionth hamburger. Billionth is an ordinal number.

2. Your answer may be different from mine. You might think 8 is a small cardinal number. Actually, 8 is a small cardinal number compared with a billion. But 3 is even smaller. Smaller yet is 1. The smallest cardinal number is zero.

3. To change inches to feet, you divide by 12 because there are 12 inches in a foot.

```
     4
12)48
```

So 48" = 4'.

4. To change inches into feet, you again divide by 12. But this time there is a remainder. Twelve doesn't go into 100 evenly.

```
     8 R 4
12)100
  − 96
     4
```

where R stands for remainder

So 100" = 8' 4".

5. On page 21 I wrote, "If the Math Building *were* very tall" it would have a fourteenth floor. I didn't write that the Math Building *was* very tall. I was using the subjunctive mood. You use the subjunctive mood when you want to indicate that something is not true or is probably not true. In the subjunctive mood, you use *be* or *were* instead of *is* or *was.*

In the movie *Fiddler on the Roof*, the hero sings, "If I were a rich man. . . ."

In actuality, the Math Building has only three floors. A cardinal number.

6. Fred's office is on the third floor. An ordinal number.

7. To go from 3' to 17' would be an increase of 14'. $17 - 3 = 14$.

8. To go from \$7 to \$1,000,000,000, we again need to subtract.

$$\begin{array}{r} 1{,}000{,}000{,}000 \\ -\qquad\qquad 7 \\ \hline \$999{,}999{,}993 \end{array}$$

9. \$999,999,993 in words: Nine hundred ninety-nine million, nine hundred ninety-nine thousand, nine hundred ninety-three dollars.

from page 27

1. Since a diameter is twice as long as a radius, we double the 17":

$$\begin{array}{r} 17 \\ \times\ 2 \\ \hline 34 \end{array}$$

A 34" pizza is a huge pizza.

2. To convert inches into feet, you divide by 12.

$$\begin{array}{r} 2 \text{ R } 10 \\ 12\overline{)\,34} \\ -\,24 \\ \hline 10 \end{array}$$

34" = 2' 10"

3. 62,804 is a cardinal number.

4. One-third of 150 = $\frac{1}{3}$ of 150 = $3\overline{)\,150}$ with quotient 50

5. Almost everyone says that x will be 5 since $4 < 5 < 6$. But there are other answers! For example, x can be 4½ since $4 < 4\frac{1}{2} < 6$.

from page 30

1.

6	clothes
0	housing
26	food
50	Sunday school offering
30	books
+ 2	miscellaneous
114	a total of $114 spent each month

2. If he makes $500 each month and spends $114, we need to subtract to find out how much he saves.

$$\begin{array}{r} 500 \\ -\ 114 \\ \hline 386 \end{array}$$

He saves $386 each month.

3. Half of $500 = ½ × 500 = $2\overset{\displaystyle 250}{\overline{)\,500}}$

4. Half of his income is $250 (from problem 3).
He saves $386 (from problem 2).
He saves more than half of his income. 386 > 250

5. If he saves $386 each month, he saves twelve times as much each year.

$$\begin{array}{r} 386 \\ \times\ \ 12 \\ \hline 772 \\ 386\ \ \\ \hline 4632 \end{array}$$

He saves $4,632 each year.

6. From the previous problem, we learned he saves $4,632 each year. The easy way to multiply a whole number by 10 is to just add a zero on the right.

For example: 88 × 10 = 880 259 × 10 = 2590 710 × 10 = 7100
So Fred would save $46,320 in ten years.

7. $463,200. Multiplying by 100 is the same as multiplying by 10 twice.

from page 39

1. The first time he doubles the price it will be $3,200.
   The second time it will be $6,400.
   The third time it will be $12,800.
   Fourth time $25,600.
   Fifth time $51,200.
   Sixth time $102,400.
   Seventh time $204,800.
   Eighth time $409,600.
   Ninth time $819,200.
   And by the tenth double, we cross a million $1,638,400.

2. If he saved $386 each month, in four months he would save four times that much.

$$\begin{array}{r} 386 \\ \times \quad 4 \\ \hline 1544 \end{array}$$

He could save $1,544.

That's not enough.

3. If he saved $386 each month, in five months he would save five times that much.

$$\begin{array}{r} 386 \\ \times \quad 5 \\ \hline 1930 \end{array}$$

He could save $1,930. That is enough to buy the $1,600 bike.

4. Since a diameter is twice as large as a radius (in the same circle), the diameter would be 42".

from page 43

1. So, if Fred spent equal amounts of time riding his bike to each of these six places, it would look like:

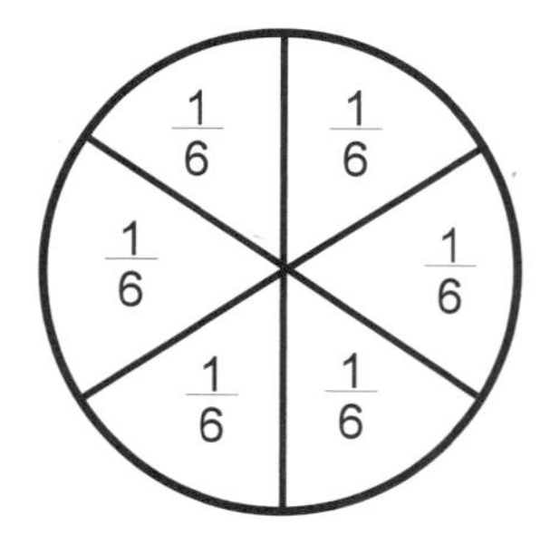

One-sixth of his time riding to class
One-sixth of his time riding to the library
One-sixth of his time riding to the park
One-sixth of his time riding to Betty's
One-sixth of his time riding to Alexander's
One-sixth of his time riding to PieOne pizza

2. 1,000,000,000 + 4,000,000,000 = 5,000,000,000 (five billion).

3. There are 60 minutes in an hour.

$$4 \text{ hours} \times 60 \text{ minutes/hour} = 240 \text{ minutes.}$$

(*60 minutes/hour* is pronounced "sixty minutes per hour.")

4. In the previous problem (problem 3) we found out that 4 hours = 240 minutes. We need one-sixth of 240 minutes.

$$\frac{1}{6} \times 240 \qquad \begin{array}{r} 40 \\ 6\overline{)240} \end{array} = \text{forty minutes}$$

5. Your answer may be different from mine. When I think of sectors, I think of pizza. Some people think of a slice of boysenberry pie.

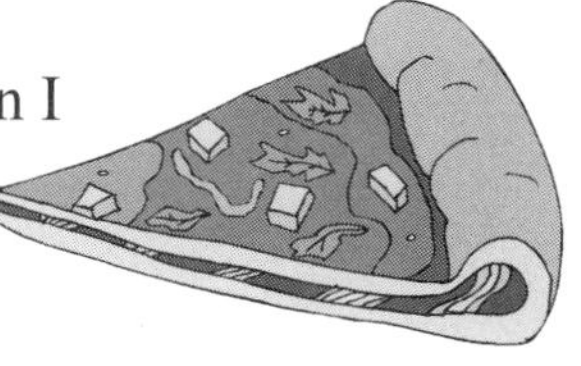

6. Any number times zero is always equal to zero.

7. If $x < 36''$ then Fred would hit his head, so we prefer $x > 36''$. Actually, $x = 36''$ would also be okay. Then his head would barely touch the underside of his desk. If you want to write that $x > 36''$ or $x = 36''$ you can write $x \geq 36''$. (x is *greater than or equal to* 36".)

8. $z \leq 55$

from page 46

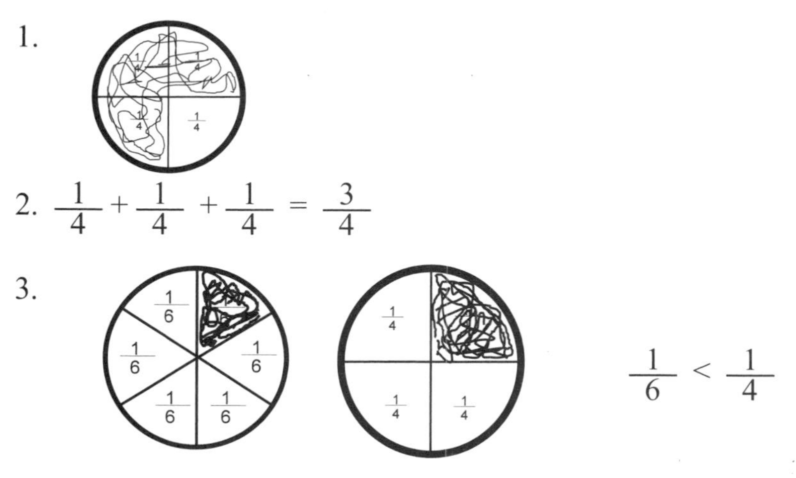

4. He had nothing left. Zero dollars. \$1,935.06 – \$1,935.06 = \$0.00

5.
$$\begin{array}{r} 4026 \\ -\ \ \ 39 \\ \hline 3987 \end{array}$$
She would have \$3,987 left.

6. Did you draw some circles to see the answer?

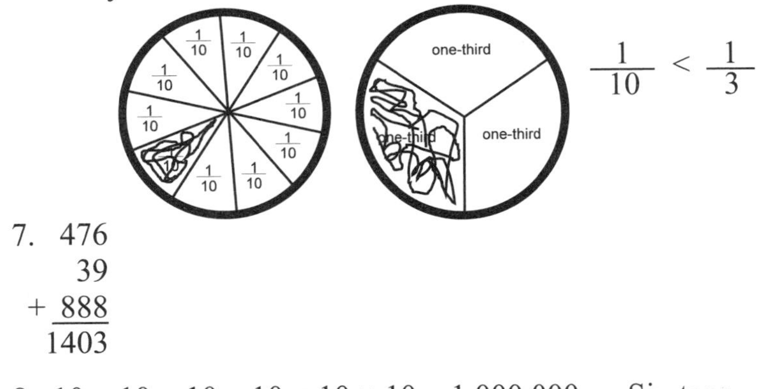

$\frac{1}{10} < \frac{1}{3}$

7.
$$\begin{array}{r} 476 \\ 39 \\ +\ \underline{888} \\ 1403 \end{array}$$

8. $10 \times 10 \times 10 \times 10 \times 10 \times 10 = 1{,}000{,}000$ Six tens.
Each ten adds another zero. A million has six zeros.

9. $\frac{1}{6} + \frac{1}{6} + \frac{2}{6} = \frac{4}{6}$

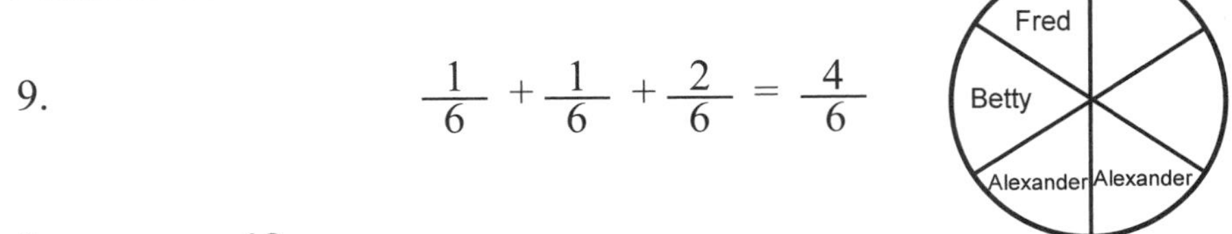

from page 48

1. Since there are six equal pieces, one piece is one-sixth of the whole comb. $\frac{1}{6}$

2. If we take three pieces, we would have half of the comb. Three pieces equals $\frac{3}{6}$ but it also equals $\frac{1}{2}$

$\frac{3}{6} = \frac{1}{2}$

3. There are two possible answers. You could have said that Betty and Alexander each get $\frac{3}{6}$ of the pencils, or you could have said that each of them get $\frac{1}{2}$ of the pencils.

Betty's Alexander's

4. You could say that each gets $\frac{2}{6}$ of the pencils, or each gets $\frac{1}{3}$

$\frac{2}{6} = \frac{1}{3}$

Fred's Betty's Alexander's

5. Each would get ten paper clips which is $\frac{10}{30}$ of all the paper clips. Or you could have said that each gets $\frac{1}{3}$ of all the paper clips.

6. $\frac{7}{21} = \frac{1}{3}$ since $\frac{7 \div 7}{21 \div 7} = \frac{1}{3}$

from page 51

1. Fred was 37 pounds and the knife was 13 pounds. $37 + 13 = 50$ pounds.

2. Fred's weight was 37 pounds out of a total of 50 pounds: $\frac{37}{50}$

3. $\frac{13}{50}$

4. Six pages ago we added fractions: $\frac{1}{6} + \frac{1}{6} + \frac{2}{6} = \frac{4}{6}$

When the bottoms of the fractions are the same, you just add the tops.

$$\frac{37}{50} + \frac{13}{50} = \frac{50}{50}$$

5. $\frac{15 \div 5}{20 \div 5} = \frac{3}{4}$

6. $\frac{1}{8} + \frac{1}{8} = \frac{2}{8}$ and then reducing the answer $\frac{2}{8} = \frac{1}{4}$

$$\frac{2}{9} + \frac{1}{9} = \frac{3}{9} = \frac{1}{3}$$

7. 93,000,000 miles × 8 furlongs/mile = 744,000,000 furlongs

from page 59

1. $\frac{4}{4} - \frac{3}{4} = \frac{1}{4}$ That's why Alexander said, "He's lost a quarter of all his blood!"

Subtraction of fractions works the same way that addition of fractions works. Look at these addition and subtraction examples and you'll get the idea:

$$\frac{2}{7} + \frac{3}{7} = \frac{5}{7}$$
$$\frac{5}{9} + \frac{2}{9} = \frac{7}{9}$$

☜ Addition

Subtraction ☞

$$\frac{8}{11} - \frac{1}{11} = \frac{7}{11}$$
$$\frac{6}{8} - \frac{4}{8} = \frac{2}{8}$$

2. $\frac{18}{21} - \frac{14}{21} = \frac{4}{21}$

3. $\frac{18}{21} - \frac{15}{21} = \frac{3}{21}$ which reduces to $\frac{1}{7}$

4.

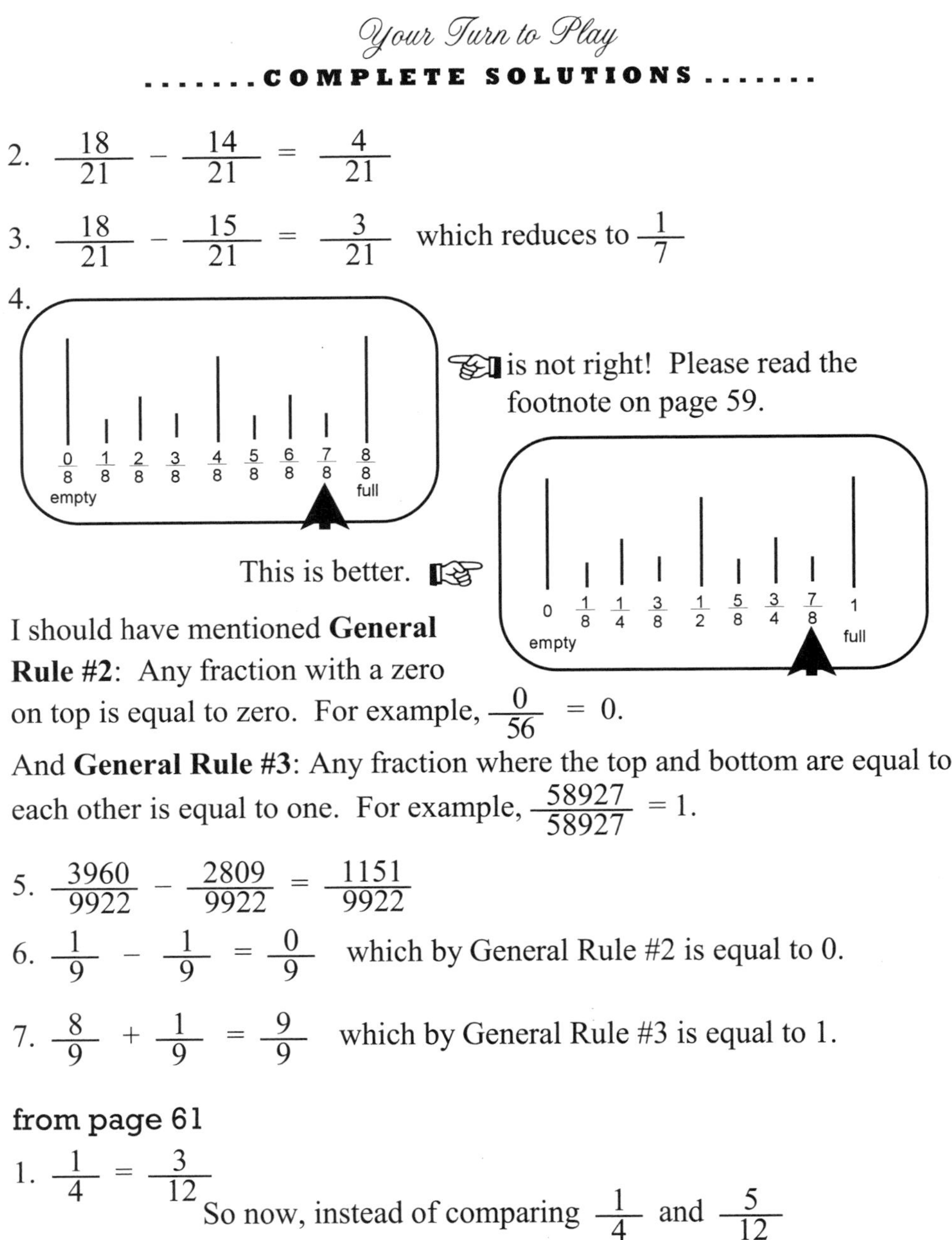

☜ is not right! Please read the footnote on page 59.

This is better. ☞

I should have mentioned **General Rule #2**: Any fraction with a zero on top is equal to zero. For example, $\frac{0}{56} = 0$.

And **General Rule #3**: Any fraction where the top and bottom are equal to each other is equal to one. For example, $\frac{58927}{58927} = 1$.

5. $\frac{3960}{9922} - \frac{2809}{9922} = \frac{1151}{9922}$

6. $\frac{1}{9} - \frac{1}{9} = \frac{0}{9}$ which by General Rule #2 is equal to 0.

7. $\frac{8}{9} + \frac{1}{9} = \frac{9}{9}$ which by General Rule #3 is equal to 1.

from page 61

1. $\frac{1}{4} = \frac{3}{12}$

So now, instead of comparing $\frac{1}{4}$ and $\frac{5}{12}$

we are comparing $\frac{3}{12}$ and $\frac{5}{12}$

so we can say $\frac{1}{4} < \frac{5}{12}$

2. In order to compare them, we need to have a common denominator. We change the $\frac{2}{3}$ into $\frac{8}{12}$ (by multiplying top and bottom by 4).

So now instead of comparing $\frac{2}{3}$ and $\frac{7}{12}$

we are comparing $\frac{8}{12}$ and $\frac{7}{12}$

So we can say $\frac{2}{3} > \frac{7}{12}$

3. You are going to have to mess with *both* fractions in order to get a common denominator. If you start with a fraction with a 5 on the bottom and another fraction with 8 on the bottom, you change them both to fractions with 40 on the bottom.

$\frac{2}{5}$ becomes $\frac{16}{40}$

$\frac{3}{8}$ becomes $\frac{15}{40}$ So $\frac{3}{8} < \frac{2}{5}$

4. $\frac{8}{8} - \frac{5}{8} = \frac{3}{8}$ He has lost three-eighths of his blood.

5. 84 – 65 = 19. It is nineteen beats/minute faster.

6. The trick is to make all the numbers have the same units. It's hard to compare minutes with hours (or feet with furlongs). Since one hour equals 60 minutes, the problem becomes *how much do we have to add to 55 minutes to get 60 minutes.* That's easy. Five minutes.

from page 66

1. XI, XII, XIII, XIV, XV, XVI, XVII, XVIII, XIX, XX
2. 60 = LX, 90 = XC, 400 = CD
3. DCLXVI
4. MCMXCIX "MCM" = 1900 "XC" = 90 "IX" = 9
5. If I were asked to divide DXLVI by XIV, I would first convert each of those numbers to Arabic numerals. Then I would do the division. Then I would convert my answer back into Roman numerals.
6. "Divide DXLVI by XIV" would become divide "546 by 14."

```
     39
14)546
  - 42
    126
  - 126
```

Then convert 39 back into Roman numerals: 39 = XXXIX

from page 70

1. When the denominators are the same, you can just add the numerators.

$\frac{1}{5} + \frac{3}{5} = \frac{4}{5}$ He had four-fifths of a tablespoon of pepper.

2. $\frac{1}{4}$ and $\frac{1}{8}$ have different denominators. The first step is to make the denominators the same. ("Find a common denominator" is how it's said.)

$\frac{1}{4}$ becomes $\frac{2}{8}$ (by multiplying top and bottom by 2).

$$\frac{1}{4} + \frac{1}{8}$$

$$\frac{2}{8} + \frac{1}{8} = \frac{3}{8}$$

3. $\frac{1}{3} + \frac{1}{4} + \frac{1}{6}$ The common denominator is 12 since 3, 4, and 6 all divide evenly into 12.

$\frac{4}{12} + \frac{3}{12} + \frac{2}{12} = \frac{9}{12}$ which reduces to $\frac{3}{4}$ (dividing top and bottom by 3.)

4. In order to compare fractions, you need to have a common denominator. $\frac{4}{5}$ becomes $\frac{16}{20}$

$\frac{3}{4}$ becomes $\frac{15}{20}$ $\frac{3}{4} < \frac{4}{5}$

5. Use before 2004.

6. $\frac{10}{60}$ which reduces to $\frac{1}{6}$

7. 1,000,000th is an ordinal number.

from page 73

1. A whole pizza is $\frac{5}{5}$ so to find out how much was left for Darlene we subtract $\frac{3}{5}$ from $\frac{5}{5}$

$$\frac{5}{5} - \frac{3}{5} = \frac{2}{5}$$

2. Three pizzas divided equally among four people: $\frac{3}{4}$ of a pizza each.

3. With pizzas, one-half plus one-half equals one whole pizza. But if you cut a tapir in half, and you try to add one-half tapir plus one-half tapir, all you will get is one dead tapir. $\frac{1}{2} + \frac{1}{2}$ doesn't equal one for tapirs.

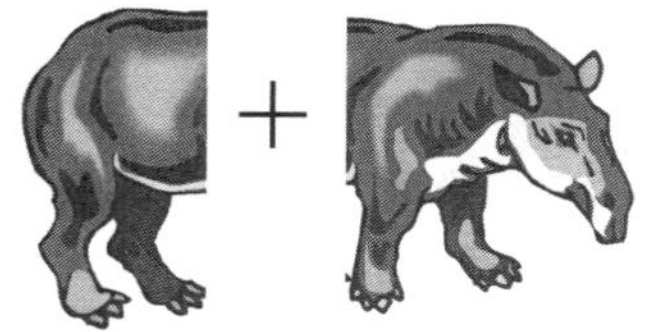

Besides, tapirs don't taste like pizza.

4. Your answers will differ from mine.
Gallons of gasoline. Quarts of ice cream. Pounds of ants. (If you are in the right store, you can say, "I'd like to buy one-third of a pound of ants, please.") Tons of sand. Minutes spent in the dentist chair. Moles of sodium. (That's chemistry talk.) Inches of rain. Acres of lawn. Miles per hour (mph) of my camel. Calories expended while boxing.

What doesn't work are flavors of ice cream. Ever have half a flavor?

from page 80

1. The least common multiple of 8 and 10 is 40.

2. The smallest number that 2, 3, 4, 5, and 6 divide into is 60.

3. That's easy. It's 256.

4. 81.

5. To add fractions, you first find the LCM of the denominators.
Or you could have written: To add fractions, you first find the least common multiple of the denominators.

6. Look at problems 3 and 4. The LCM wasn't greater than each of the numbers. In problem 3, for example, the LCM wasn't greater than 256. What is true is that the LCM will be $\geq$ to each of the numbers.

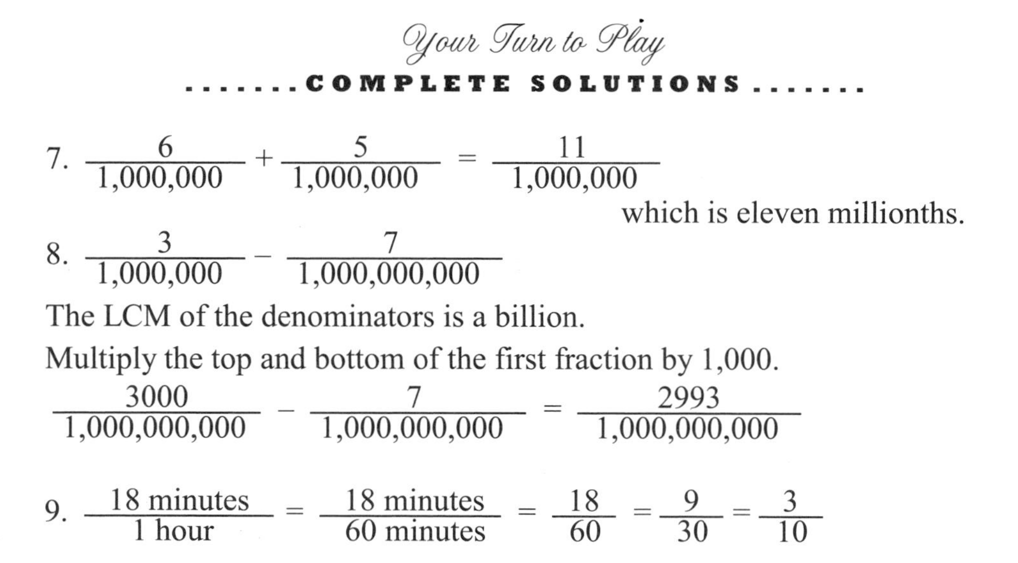

7. $\dfrac{6}{1{,}000{,}000} + \dfrac{5}{1{,}000{,}000} = \dfrac{11}{1{,}000{,}000}$

which is eleven millionths.

8. $\dfrac{3}{1{,}000{,}000} - \dfrac{7}{1{,}000{,}000{,}000}$

The LCM of the denominators is a billion.

Multiply the top and bottom of the first fraction by 1,000.

$\dfrac{3000}{1{,}000{,}000{,}000} - \dfrac{7}{1{,}000{,}000{,}000} = \dfrac{2993}{1{,}000{,}000{,}000}$

9. $\dfrac{18 \text{ minutes}}{1 \text{ hour}} = \dfrac{18 \text{ minutes}}{60 \text{ minutes}} = \dfrac{18}{60} = \dfrac{9}{30} = \dfrac{3}{10}$

Eighteen minutes is three-tenths of an hour.

from page 83

1. LCD is the standard abbreviation for least common denominator.

(LCD is also the standard abbreviation for liquid-crystal display which you see on some watches and calculators.)

2. In order to add fractions, you need to find the LCD of the fractions—which is the LCM of the denominators.

3.
```
    13 R 3
12) 159
   -12
    39
   -36
     3
```
159" = 13' 3"

4. 159" = $13\frac{3}{12} = 13\frac{1}{4}$ feet

5.
```
   14 R 2
4) 58
  -4
   18
  -16
    2
```
58 quarts = 14 gallons 2 quarts

6. 58 quarts $= 14\frac{2}{4} = 14\frac{1}{2}$ gallons

from page 86

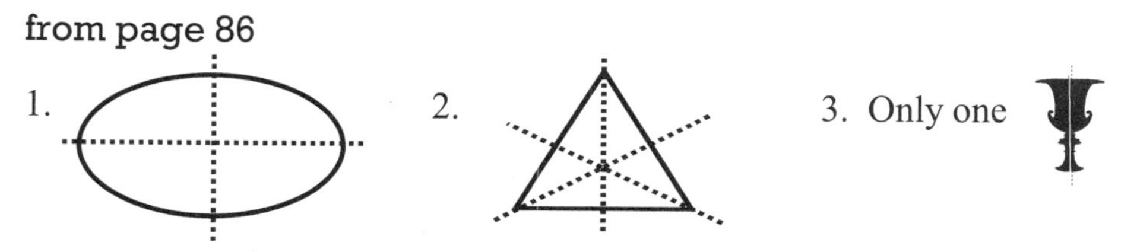

4. I can't. The handle on the pitcher messes things up.

5. 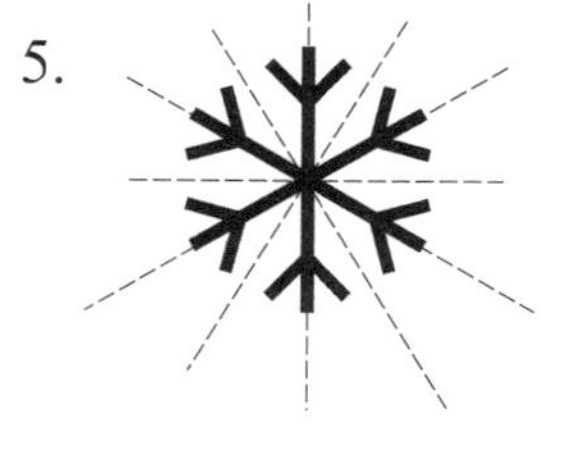

6. There are many possible answers. I'll draw one possible answer on page 90. If I drew it here, you might accidentally see it and that would spoil your chance to discover it on your own.

from page 89

1. Zero times what equals 14? That's a really hard question. Zero times six is zero. Zero times 297923923423779995 is zero. Zero times zero is zero. Zero times anything is always zero. There is no number that works. The question "$0 \times ? = 14$" has no answer.

2. That's easy. $3 + 3 + 3 + 3 + 3 + 3 + 3 = 21$. You add up seven 3s to get 21.

3. $\frac{21}{3}$ means *How many 3s do you have to add up to get 21?*

4. $\frac{20}{5}$ means *How many 5s do you have to add up to get 20?*

   $5 + 5 + 5 + 5 = 20$. I used four 5s. So $\frac{20}{5} = 4$.

5. $\frac{6}{1}$ means *How many 1s do you have to add up to get 6?*

   $1 + 1 + 1 + 1 + 1 + 1 = 6$. I used six 1s. So $\frac{6}{1} = 6$.

6. $\frac{14}{0}$ asks the silly question *How many 0s do you add up to get 14?*

I'll start adding and you tell me when to stop: 0 + 0 Hey! You never said stop.

This is why **Division by zero doesn't make sense.**

You can call that General Rule #4 if you like.

from page 97

1. In order to subtract fractions, they need to have the same denominators. (The procedures for addition and subtraction are almost alike.)
The LCD for $\frac{7}{8}$ and $\frac{1}{4}$ is 8.

We change $\frac{1}{4}$ into $\frac{2}{8}$ by multiplying top and bottom by 2.

$$\frac{7}{8} - \frac{1}{4} = \frac{7}{8} - \frac{2}{8} = \frac{5}{8}$$

2. The LCD for $\frac{1}{3}$ and $\frac{1}{4}$ is 12. $\frac{1}{3} = \frac{4}{12}$ and $\frac{1}{4} = \frac{3}{12}$

$$\frac{1}{3} - \frac{1}{4} = \frac{4}{12} - \frac{3}{12} = \frac{1}{12}$$

3. $\frac{7}{8} - \frac{1}{5} = \frac{35}{40} - \frac{8}{40} = \frac{27}{40}$ lb.

4. $\frac{1}{2} - \frac{1}{3} = \frac{3}{6} - \frac{2}{6} = \frac{1}{6}$ hour

5. One-sixth of an hour $= \frac{1}{6}$ of 60 minutes $= 6\overline{)60}$ with quotient 10 $= 10$ minutes.

6. One-sixth of \$12 $= \frac{1}{6} \times \$12 = \$2$.

7. 17 jumps × 4 hoof marks/jump = 68 hoof marks.

8. They are all disappearing at the same time. It would take three days for all of them to go away.

9. Do you remember how 3 pizzas divided by 8 students gave each of them $\frac{3}{8}$ of a pizza? In the case of the lamb shoes, 3 pounds divided into 4 equal parts, means that each shoe weighs $\frac{3}{4}$ lb.

10. A whole bait basket is $\frac{17}{17}$ and Joe ate $\frac{6}{17}$ of it.
$\frac{17}{17} - \frac{6}{17} = \frac{11}{17}$ He left eleven-seventeenths of the basket uneaten.

Your Turn to Play
.......COMPLETE SOLUTIONS.......

from page 99

1. $\frac{1}{3} \times \frac{3}{4} = \frac{3}{12}$

If you left your answer as $\frac{3}{12}$ it would be marked wrong.

Here are the General Rules. I'll put them in a box for handy reference.

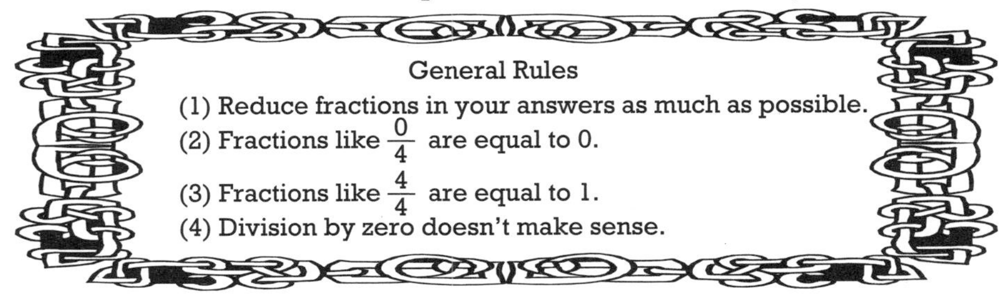

General Rules

(1) Reduce fractions in your answers as much as possible.
(2) Fractions like $\frac{0}{4}$ are equal to 0.
(3) Fractions like $\frac{4}{4}$ are equal to 1.
(4) Division by zero doesn't make sense.

So the correct solution would be $\frac{1}{3} \times \frac{3}{4} = \frac{3}{12} = \frac{1}{4}$

from page 101

1. $5\frac{1}{6} = \frac{5 \times 6}{1 \times 6} + \frac{1}{6} = \frac{30}{6} + \frac{1}{6} = \frac{31}{6}$
2. $11\frac{1}{10} = \frac{11 \times 10}{1 \times 10} + \frac{1}{10} = \frac{110}{10} + \frac{1}{10} = \frac{111}{10}$
3. $2\frac{5}{8} = \frac{2 \times 8}{1 \times 8} + \frac{5}{8} = \frac{16}{8} + \frac{5}{8} = \frac{21}{8}$

from page 101

1. $5\frac{1}{6} = \frac{31}{6}$ *6 times 5 . . . plus 1*
2. $11\frac{1}{10} = \frac{111}{10}$ *10 times 11 . . . plus 1*
3. $2\frac{5}{8} = \frac{21}{8}$ *8 times 2 . . . plus 5*

from page 102

1. $14 \times 3\frac{1}{7} = \frac{14}{1} \times \frac{22}{7} = \frac{308}{7} = 44''$

from page 104

1. Yes. We now have the commutative law of addition and the commutative law of multiplication.

2. $\frac{3}{7} - \frac{1}{8} = \frac{24}{56} - \frac{7}{56} = \frac{17}{56}$

3. It's very rare that a – b = b – a. (It only happens when a and b are the same number.) So subtraction is not commutative.

4. D < M means 500 < 1000. It is true.

5. $8\frac{1}{4} \times 3\frac{1}{7} = \frac{33}{4} \times \frac{22}{7} = \frac{726}{28} = 28\overline{)726}$ = 25 R 26

$= 25\frac{26}{28} = 25\frac{13}{14}$

6. Is it always true that a ÷ b equals b ÷ a? That's the same as asking if it is always true that $\frac{a}{b} = \frac{b}{a}$ Division is not commutative.

7. It sure would be difficult to put your shoes on first. On the other hand, I understand what *commutative* means, and I have told my daughters when they were young to "Go and put on your shoes and socks." It's just an expression in English. When I told them to "make your beds," I didn't expect them to get out hammers and build a couple of beds.

from page 107

1.
$$3\frac{3}{4} = 3\frac{3}{4}$$
$$+\ 2\frac{1}{2} = 2\frac{2}{4}$$
$$5\frac{5}{4} = 5 + 1\frac{1}{4} = 6\frac{1}{4}$$

2.
$$1\frac{1}{2} = 1\frac{3}{6}$$
$$2\frac{1}{3} = 2\frac{2}{6}$$
$$+\ 4\frac{2}{3} = 4\frac{4}{6}$$
$$7\frac{9}{6} = 7 + 1\frac{3}{6} = 8\frac{1}{2}\text{ oz.}$$

3. $89\frac{1}{7} + 2\frac{1}{7} = 91\frac{2}{7}$

4. $89\frac{1}{7} \times 2\frac{1}{7} = \frac{624}{7} \times \frac{15}{7} = \frac{9360}{49} = 49\overline{)9360}$ = 191 R 1 $= 191\frac{1}{49}$

$$\begin{array}{r} 191 \text{ R } 1 \\ 49\overline{)9360} \\ -49 \\ \hline 446 \\ -441 \\ \hline 50 \\ -49 \\ \hline 1 \end{array}$$

5. $89\frac{1}{7} - 2\frac{1}{7} = 87$

6. Anything times zero is always equal to zero.

7. Anything times zero is always equal to zero.

8. $89\frac{1}{7} + 0 = 89\frac{1}{7}$

9. The LCD (least common denominator) of $\frac{2}{5}$ and $\frac{3}{7}$ is 35.

   $\frac{2}{5} = \frac{14}{35}$ $\frac{3}{7} = \frac{15}{35}$ Therefore, $\frac{2}{5} < \frac{3}{7}$

10. 42 pennies = 42¢. 27 nickels = 27 × 5 = 135¢.

    3 dimes = 3 × 10 = 30¢. 6 quarters = 6 × 25 = 150¢.

    42¢ + 135¢ + 30¢ + 150¢ = 357¢

from page 114

1. One-tenth of $1\frac{1}{4} = \frac{1}{10} \times 1\frac{1}{4} = \frac{1}{10} \times \frac{5}{4} = \frac{5}{40} = \frac{1}{8}$ oz.

2. $\frac{1}{9} \times \frac{6}{7} = \frac{1}{\cancel{9}_3} \times \frac{\cancel{6}^2}{7} = \frac{2}{21}$

3. $\frac{5}{6} \times \frac{4}{15} = \frac{\cancel{5}^1}{\cancel{6}_3} \times \frac{\cancel{4}^2}{\cancel{15}_3} = \frac{2}{9}$

4. $\frac{1}{7} + \frac{1}{77} = \frac{11}{77} + \frac{1}{77} = \frac{12}{77}$

5. $\frac{1}{7} - \frac{1}{77} = \frac{11}{77} - \frac{1}{77} = \frac{10}{77}$

6. $3\frac{3}{5} \times \frac{5}{6} = \frac{\cancel{18}^3}{\cancel{5}_1} \times \frac{\cancel{5}^1}{\cancel{6}_1} = \frac{3}{1} = 3$

7. One-seventh of 1813 $= \frac{1}{7} \times 1813 = \frac{1}{7} \times \frac{1813}{1} = \frac{1813}{7} =$

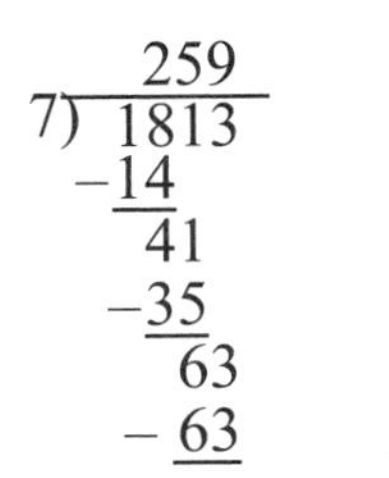

Some readers who really like English may have noticed that the word "of" sometimes gets translated as "times."

One-seventh of 1813
$\frac{1}{7} \times 1813$

from page 116

1. The opposite of "adding six to a number" is "subtracting six from a number." If I first add six, and then I subtract six, I end up where I started.

Suppose I start with the number 40. Here's how it would look:

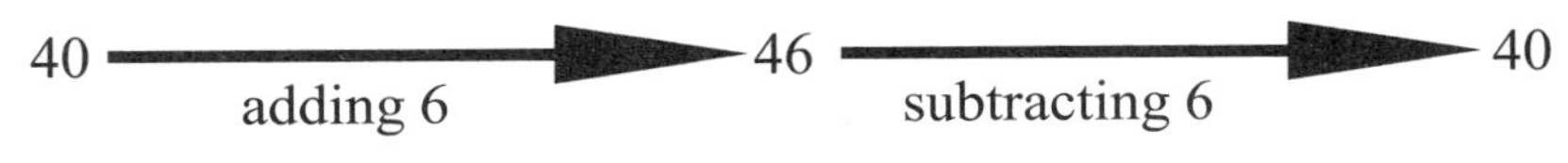

2. The opposite of "multiplying by seven" is "dividing by seven."

Suppose I start with the number 91. Here's how it would look:

91 → (multiplying by 7) → 637 → (dividing by 7) → 91

3. It is "going west for an hour."

4. This is a function that doesn't have an opposite. "dividing by zero" won't work because it doesn't make sense.*

If I tell you I'm thinking of a number and after multiplying by zero the answer is zero, can you tell me what my original number was? You can't.

✶ That was one of the general rules:
(1) Reduce fractions in your answers as much as possible.
(2) Fractions like $\frac{0}{4}$ are equal to 0.
(3) Fractions like $\frac{4}{4}$ are equal to 1.
(4) Division by zero doesn't make sense.
(5) Canceling is only done when you're multiplying fractions.

In contrast, if I tell you I'm thinking of a number and after multiplying by seven my answer is 637, you can tell me that my original number was 91. Some functions have opposites, and some don't.

5. It would be changing an improper fraction back into a mixed number.

6. There is no way to undo that function. Washing that pizza or putting it in the freezer won't take the burned mess and turn it back into an unbaked pizza.

7. My function, "multiply by six and then add twenty-four," would take 100 and do this:

100 ⟶ 624
my function at work

The opposite function (known as the **inverse function**) is *not* "Divide by six and then subtract twenty-four." That would turn 624 into 80. (Try it!)

Instead, the inverse function to "multiply by six and then add twenty-four" is "subtract twenty-four and then divide by six."

8. $6\frac{2}{5} \times 6\frac{1}{4} = \frac{32}{5} \times \frac{25}{4} = \frac{\overset{8}{\cancel{32}}}{\underset{1}{\cancel{5}}} \times \frac{\overset{5}{\cancel{25}}}{\underset{1}{\cancel{4}}} = \frac{40}{1} = 40$

9. There are several different answers that would be correct. Many readers will say that the inverse to "multiply by one" is "divide by one." This works.

But another correct answer is "multiply by one." Suppose I start with the number 955. Here's how it would look:

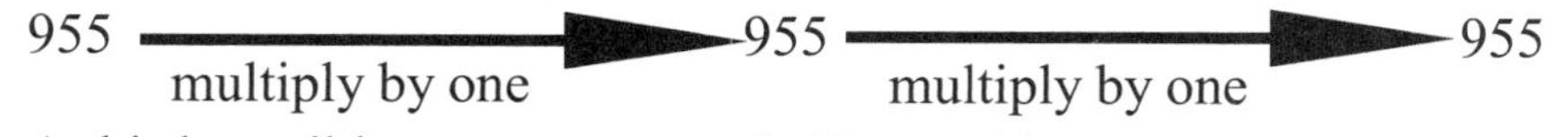

A third possible correct answer is "Do nothing." Suppose I start with the number 345. Here's how it would look:

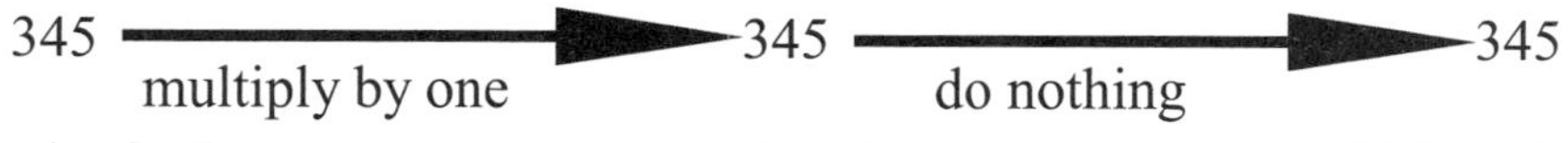

The further you go in mathematics, the more often you will find that there are several ways to do something.

After arithmetic comes algebra.

After algebra comes geometry.

In geometry, for example, there are (at least) four different ways to prove *If two sides of a triangle are equal in length, then the opposite angles are equal.*

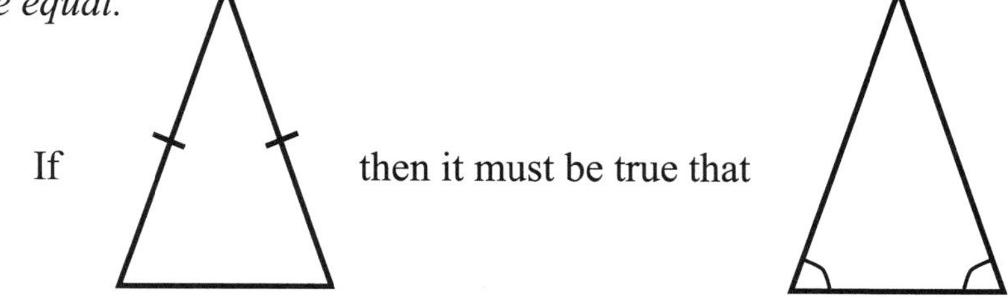

10. $2\frac{5}{8} \times 2\frac{5}{8} = \frac{21}{8} \times \frac{21}{8} = \frac{441}{64} = 64\overline{)441}$ = 6 R 57; $441 - 384 = 57$; $= 6\frac{57}{64}$

from page 119

1. $9\frac{3}{4} \times 14\frac{1}{3} = \frac{39}{4} \times \frac{43}{3} = \frac{\cancel{39}^{13}}{4} \times \frac{43}{\cancel{3}_1} = \frac{559}{4} = 4\overline{)559}$ = 139 R 3

$= 139\frac{3}{4}$ square inches.

2. $3\overline{)160}$ = 53 R 1 $= 53\frac{1}{3}$ yards

3. $\ell = 120$ yards. $w = 53\frac{1}{3}$ yards. Area $= 120 \times 53\frac{1}{3} = \frac{120}{1} \times \frac{160}{3}$
$= \frac{\cancel{120}^{40}}{1} \times \frac{160}{\cancel{3}} = \frac{6400}{1} = 6400$ square yards.

4. High School:
$23'\,4'' \times 10' = 23\frac{1}{3} \times 10 = \frac{70}{3} \times \frac{10}{1} = \frac{700}{3} = 233\frac{1}{3}$ sq. ft.
College:
$18'\,6'' \times 20' = 18\frac{1}{2} \times 20 = \frac{37}{2} \times \frac{20}{1} = \frac{740}{2} = 370$ sq. ft.

from page 122

1. 4 hours × 60 mph $= \frac{4 \text{ hrs.}}{1} \times \frac{60 \text{ miles}}{\text{hr.}} = \frac{4 \cancel{\text{hrs.}}}{1} \times \frac{60 \text{ miles}}{\cancel{\text{hr.}}}$
$= 240$ miles

2. 43 gallons $\times \frac{4 \text{ quarts}}{1 \text{ gallon}} = \frac{43 \cancel{\text{gallons}}}{1} \times \frac{4 \text{ quarts}}{1 \cancel{\text{gallon}}} = 172$ quarts

3. 888 $\cancel{\text{inches}} \times \frac{1 \text{ ft.}}{12 \cancel{\text{inches}}} = 12\overline{)888}$ with quotient 74 $= 74$ feet.

One common mistake is to write the conversion factor upside down.

If you wrote 888 inches $\times \frac{12 \text{ inches}}{1 \text{ ft}}$ you would see that it wouldn't work because you couldn't cancel the inches.

4. Area $= 3\frac{1}{3} \times 2\frac{2}{5} = \frac{10}{3} \times \frac{12}{5} = \frac{\cancel{10}^{2}}{\cancel{3}_{1}} \times \frac{\cancel{12}^{4}}{\cancel{5}_{1}} = \frac{8}{1} = 8$ sq. ft.

5. The square of 12 $= 12 \times 12 = 144$

6. The square of 12 is 144. The square root of 144 is 12. The inverse function takes you back to where you started.

If I square 5, I get 25. The square root of 25 is 5.

7. The square root of 100 is 10 (since $10 \times 10 = 100$).

8. The square root of 81 is 9.

9. If I take the square root of 36, I get 6. How do you get back to 36? Easy. You square it. Squaring a number is the inverse function to taking the square root.

from page 125

1. $\begin{array}{r} 30 \\ -\ 5\frac{1}{7} \\ \hline \end{array}$ $\qquad$ $\begin{array}{r} 29\frac{7}{7} \\ -\ 5\frac{1}{7} \\ \hline 24\frac{6}{7} \end{array}$

2. $\begin{array}{r} 200 \\ -\ \frac{5}{6} \\ \hline \end{array}$ $\qquad$ $\begin{array}{r} 199\frac{6}{6} \\ -\ \frac{5}{6} \\ \hline 199\frac{1}{6} \end{array}$

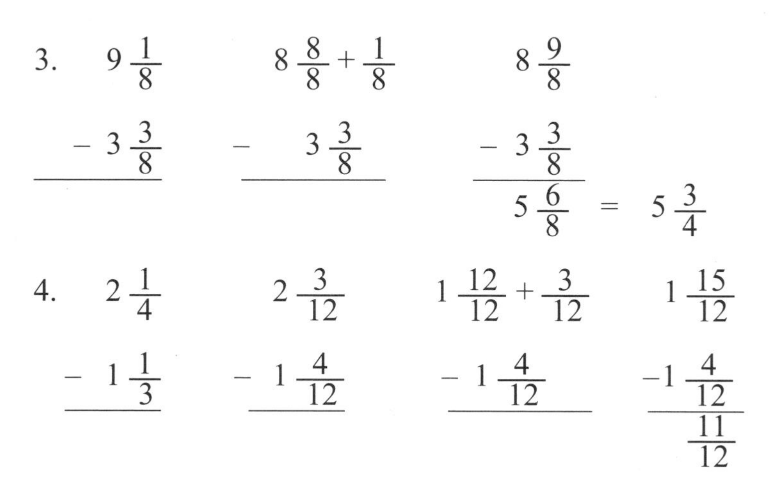

3. $9\frac{1}{8} - 3\frac{3}{8}$ $\quad$ $8\frac{8}{8}+\frac{1}{8} - 3\frac{3}{8}$ $\quad$ $8\frac{9}{8} - 3\frac{3}{8} = 5\frac{6}{8} = 5\frac{3}{4}$

4. $2\frac{1}{4} - 1\frac{1}{3}$ $\quad$ $2\frac{3}{12} - 1\frac{4}{12}$ $\quad$ $1\frac{12}{12}+\frac{3}{12} - 1\frac{4}{12}$ $\quad$ $1\frac{15}{12} - 1\frac{4}{12} = \frac{11}{12}$

from page 136

1. $10\frac{1}{10} \div 6\frac{1}{5} = \frac{101}{10} \div \frac{31}{5} = \frac{101}{10} \times \frac{5}{31} = \frac{101}{\cancel{10}_{2}} \times \frac{\cancel{5}^{1}}{31}$
$= \frac{101}{62} = 1\frac{39}{62}$

2. $\frac{1}{6} \div \frac{1}{4} = \frac{1}{6} \times \frac{4}{1} = \frac{2}{3}$ (I skipped a step.)

3. $36 \div 4\frac{4}{5} = \frac{36}{1} \div \frac{24}{5} = \frac{36}{1} \times \frac{5}{24} = \frac{\cancel{36}^{3}}{1} \times \frac{5}{\cancel{24}_{2}}$
$= \frac{15}{2} = 7\frac{1}{2}$ minutes

4. $36 \text{ letters} \div 4\frac{4}{5}\,\frac{\text{letters}}{\text{minutes}} = \frac{36 \text{ letters}}{1} \div \frac{24 \text{ letters}}{5 \text{ minutes}}$
$= \frac{36 \text{ letters}}{1} \times \frac{5 \text{ minutes}}{24 \text{ letters}} = \frac{\cancel{36}^{3}\,\cancel{\text{letters}}}{1} \times \frac{5 \text{ minutes}}{\cancel{24}_{2}\,\cancel{\text{letters}}} =$
$= \frac{15 \text{ minutes}}{2} = 7\frac{1}{2}$ minutes

5. $8\frac{1}{3} - 7\frac{2}{3} = 7\frac{3}{3} + \frac{1}{3} - 7\frac{2}{3} = \frac{2}{3}$

6. Since $7 \times 7 = 49$, we may say that 7 is the square root of 49.

7. You need to work either in feet or in inches. I'll do it both ways:

In feet: $12'\,3'' \times 4'\,4'' = 12\frac{1}{4} \times 4\frac{1}{3} = \frac{49}{4} \times \frac{13}{3} = \frac{637}{12} =$

$$12\overline{)637} \quad \text{53 R 1} \quad = 53\frac{1}{12} \text{ square feet.}$$

In inches: $12'\,3'' \times 4'\,4'' = 147'' \times 52'' = 7644$ square inches.

8. The integers = { . . . –3, –2, –1, 0, 1, 2, 3, . . . },
and the whole numbers = { 0, 1, 2, 3, . . . }, so an integer that is not a whole number is any negative number like –3 or –200.

9. The smallest number that both 10 and 12 evenly divide into is 60.

10. $\frac{1}{2} \div \frac{1}{4} = \frac{1}{2} \times \frac{4}{1} = \frac{4}{2} = 2$

$\frac{1}{2} \div \frac{1}{4}$ asks the question, "How many quarters are in a half?"

How many quarters of a pie would you need to make a half of a pie?

Two.

from page 139

1. The object of this question was to have you learn *how to find the truth*, not to learn the fact that in any four-sided figure, if three of the vertices (that's the plural of *vertex*) are right angles, then the fourth vertex must also be a right angle.

Much of why we study math is to *learn to think* rather than to just stuff your head with information. The only things you really have to sit down and memorize in math are the addition and multiplication tables. When someone asks, "What's seven times eight?" you should be able to answer, "Fifty-six."

Your education shouldn't be just adding facts to your brain. That's what paper is good for. Why memorize the phone book? Education should help you become smarter—to think more clearly—not just able to recite a bunch of facts.

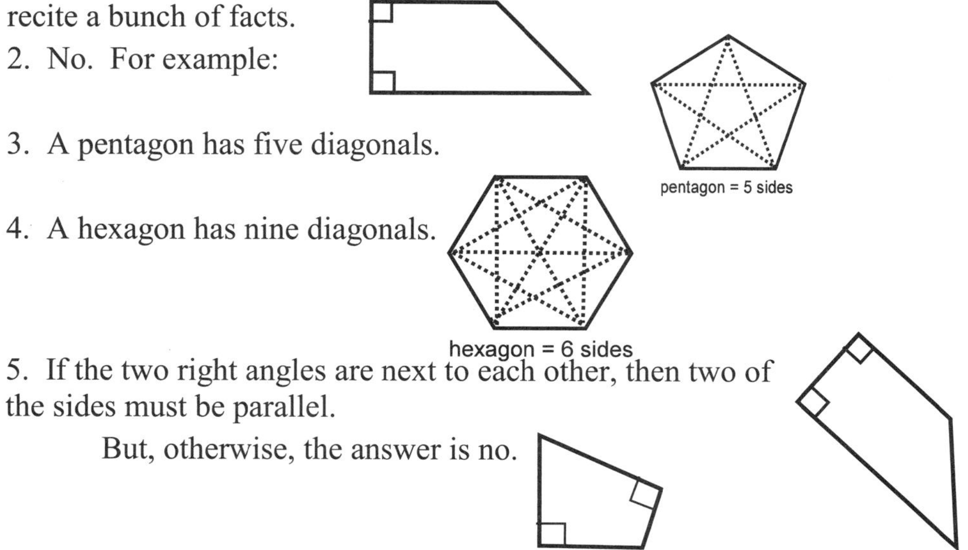

2. No. For example:

3. A pentagon has five diagonals.

4. A hexagon has nine diagonals.

5. If the two right angles are next to each other, then two of the sides must be parallel.

But, otherwise, the answer is no.

6. 19,700 diagonals. I didn't get the answer by drawing a 200-sided figure and counting all the diagonals! Here is how I arrived at 19,700. First, there are 197 diagonals that can be drawn from each vertex. (Look at the hexagon in problem 4. I can draw 3 diagonals from each vertex.) Second, there are 200 vertices. So I could draw $197 \times 200 = 39{,}400$ diagonals. Third, each diagonal would be drawn *twice*—once from each end, so I need to divide by two. $39{,}400 \div 2 = 19{,}700$.

from page 141

1. There were 4 groups of parts and each group had about 10 pieces in it, so there are approximately $4 \times 10 = 40$ pieces.

2. You have six bags, and each weighs a little more than two ounces. So you expect all six bags to weigh a little more than twelve ounces.

3. Four times 42 is roughly 4×40 which is 160. So the number of buttons is a little larger than 160, which is closest to 200.

4. In A) you have two numbers that are each greater than 3. Their product must be larger than nine. In D) you have two numbers that are less than 10, so their product would be less than 100.

Index

< 15

> 17, 166

≤ 43, 171

≥ 43

alliteration 70, 84, 117

ante meridiem 44

area of a rectangle 119

basic rule about car washes 120

billion 19

billionths 80

borrowing 124

Boyle's law 137

bridges 31

calculators 7

canceling 114

canceling units 121

cardinal numbers 21

circumference 98

common denominator 79

commutative law 103

conversion factor 122

denominator 60

desert

  the Great Australian 55

  the Sahara 55

diagonals 138

diameter 26

dimensional analysis 122

dividing by fractions

  how 133

  why 134, 135

doubling 38, 39, 127

ellipse 86

equilateral triangle 86

estimating answers 141

Euripides 17

Fiddler on the Roof 168

Florence Nightingale 113

football goal posts 119

function 116

furlong 51, 55

general rules 59, 174, 180, 181

hexagon 139

hyperbole 28, 84

imaginary numbers 73

improper fraction 81, 82

integers 73, 82

inverse function 185

invert 134

King Solomon 73

LCM 80

least common denominator 81

least common multiple 80

libra = lb 96

line of symmetry 85

lying vs. *laying* 95

man named Giggles 88

meridiem 44

million 172

millionths 80

mirror image 85

mixed numbers 82

mixed numbers into improper fractions 101

multiplying fractions 98

numerals. 64
numerator. 60
oboe. 109
of sometimes means multiply. 184
onomatopoetic words. 19, 55
opposite of. 116
ordinal numbers. 21
parallel lines. 139
pentagon. 139
Prof. Eldwood's *Box Opening the Modern Way*. 137
Prof. Eldwood's *Guide to Modern Medicine*. 62
quarts into gallons. 144
radius. 26
reciprocal. 133
reducing fractions. 48
right angle. 138
Roman numerals. 65, 66
sectors. 43
Shakespeare. 17
Smokey the Bear. 105
square of a number. 122
square root. 122
subjunctive mood. 53, 168
sum. 43
tapir. 73
tortoise and the hare. 32
trapezoid. 92
uncia (Latin) = *onza* (Italian) = oz. . . 96
unit analysis. 122
vertex. 138
whole numbers. 73
Wind in the Willows. 18, 19
writing in books. 16
zoolatry. 131